MW00474091

Volume 7

MODERN FIGHTING AIRCRAFT

F/A-18

HORNET

Published by Arco Publishing, Inc.

New York

A Salamander Book

Published by Arco Publishing Inc.,
215 Park Avenue South,
New York, N.Y. 10003,
United States of America.

Library of Congress Cataloging in Publication Data
Spick, Mike.
F/A-18 Hornet.
(Modern fighting aircraft; v.7)
"A Salamander book."
1. Hornet (Jet fighter plane) I. Title. II. Title:
F/A-18 Hornet. III. Series.
UG1242.F5S64 1984 358.4'3 84-9233
ISBN 0-668-06071-9

All correspondence concerning the content of this book should be addressed to:
Salamander Books Ltd.,
Salamander House, 27 Old Gloucester Street,
London WC1N 3AF, United Kingdom.

This book may not be sold outside the United States of America and Canada.

Credits

Project Manager: Ray Bonds

Editor: Bernard Fitzsimons

Designers: David Allen, Barry Savage

Diagrams: TIGA (© Salamander Books Ltd.)

Three-view drawings, small colour profile: © Pilot Press Ltd.

Large colour profiles: Mike Trim (© Salamander Books Ltd.)

Cutaway drawing: Mike Badrocke (© Salamander Books Ltd.)

Contents

Acknowledgements

The author and editor would like to thank all those who have contributed information and pictures to this book. Photograph sources are credited individually at the end of the book, but particular thanks are due to Timothy J. Beecher and Al Gingerich of McDonnell Aircraft Company, Saint Louis; Geoffrey Norris and Karen Stubberfield of McDonnell Douglas Corporation (UK); Robert F. Dorr, who compiled the table on page 59; Robert L. Lawson of *The Hook*; Kearney Bothwell and V. Leon Levitt of Hughes Aircraft Company, Radar Systems Group; B. McCoy and John E. James of Kaiser Electronics; Anna C. Urband of the US Naval Office of Information; and J. K. Corfield of the Northrop Corporation.

Author

Mike Spick was born in London less than three weeks before the Spitfire made its maiden flight. Educated at Churchers College, Petersfield, he later entered the construction industry and carried out considerable work on RAF airfields. An interest in wargaming led him to a close study of air warfare and a highly successful first book, *Air Battles in Miniature* (Patrick Stephens, 1978). His subsequent work includes a historical study of the evolution of air combat tactics, *Fighter Pilot Tactics* (Patrick Stephens, 1983), and he is co-author of two previous Salamander books, *Modern Air Combat* (with Bill Gunston, 1983), and an earlier volume in this series, *F-4 Phantom II* (with Doug Richardson, 1984).

Jacket: Stephen Seymour

Filmset by SX Composing Ltd.

Colour reproduction by Rodney Howe Ltd.

Printed in Belgium by Henri Proost et Cie.

Introduction

Criticism of Western fighter designs has been a growth industry for more years than I care to remember, largely because of the inability of the experts to predict accurately the form that future war in the air will take. Thus the first homing missiles were expected to put an end to manoeuvring close combat and make guns redundant: in fact, whereas it had often been possible to evade a gun attack by accelerating out of range, the reach of the new weapons made such a course distinctly unwise; the ability to turn hard became more rather than less necessary, in order to make the missile work hard to catch the target and increase the chances of failure. Manoeuvre may start at longer ranges, but is more vital than ever before.

The F-18 has received more than its fair share of criticism. It is not as capable as the F-14 in the fleet defence role; it is inferior to the F-16 in close combat; it lacks the range of the A-7; and so on. Yet the Hornet has the task of replacing not one but two types in the US Navy inventory, the A-7 Corsair and the F-4 Phantom – and the latter, the most versatile and capable fighter of its generation, is a particularly hard act to follow.

It should not be forgotten that almost every fighter ever built has sooner or later been festooned with ordnance and asked to do something for which it was not originally designed. The fact that the Hornet has rather more provision than usual for weapons delivery built in from the outset must be a major point in its favour. It must also be remembered that all modern fighters are the result of a series of compromises.

The other main grounds for criticism is the fact that the F-18 is derived from the YF-17, the loser to the YF-16 in the USAF's lightweight fighter competition in the 1970s. But it is too easy to say that the Navy bought the loser: the YF-17 performed extremely creditably in the competition and in some areas was actually superior to the YF-16. The F-18 in turn is a much improved machine, and the Navy is satisfied that it is getting the best aircraft for its purposes.

Development

Speaking on June 30, 1981, Vice Admiral Wesley L. McDonald, US Navy, addressed the critics of the F/A-18 programme: "After seven years of design, development, open testing and intensive study, the capability of this aircraft is well understood. While some factions have been criticizing legitimate problems that typically are not known until much later in service, others have been fixing those problems. Today, no significant discrepancies remain . . ." While these comments certainly seem justified, the Hornet's origins can be traced back beyond those seven years to Northrop design studies of the 1960s.

The first half-century of air warfare saw fighters grow larger, faster, heavier, less manoeuvrable, much more versatile, and vastly more expensive. No nation has an unlimited defence budget, and with the increase in costs came a diminution in fleet size. The trend firmly established itself: fewer aeroplanes, but of greater capability. As combat aircraft grew more expensive, attempts were made to break out of the vicious circle of rising costs with such aircraft as the lightweight Folland Midge and its Gnat derivative. But a main argument against the concept was that the number of young men with both the desire and the ability to fly complex fast jets was not large; it made little sense to equip them with anything less than the very best. This attitude was fine for the wealthier industrial nations, but not so good for the others.

An early, and very successful example of an austere lightweight fighter was the F-5 Freedom Fighter, manufactured by the Northrop Corporation. As early as 1954, a Northrop team toured both Europe and Asia, sounding out the defence needs of many nations. Their findings led to the development of the N-156, later to become the F-5, as a private venture.

Basically, the fighter force that a nation thinks it needs and the fighter force that it can afford are two different things. Force size and quality are limited by capital outlay, maintenance costs and skilled manpower resources, while the number of fighters available for action at any one time is governed by force size and maintainability. The F-5 was designed to offer supersonic performance as a fighter and a secondary strike capability, coupled with the simplicity, reliability and maintainability to provide a high sortie rate. Its sole concession to complexity was in the use of two engines, an early example of systems redundancy, although its fuel costs were assessed as less than half those of the single-engined F-104. It first flew on July 30, 1959.

The United States Defense Department ran a Military Assistance Program (MAP) to help the smaller aligned nations acquire suitable defence equipment. On April 25, 1962, the DoD designated the F-5 as the MAP all-purpose fighter, which effectively meant that it could be supplied to friendly nations on very advantageous, heavily subsidized terms. As a result the F-5 was to see service with the air arms of more than two dozen countries.

With the F-5 an on-going project, it was time to look to the future. Northrop had no way of knowing how long the F-5 would stay in production, and that 20 years later it would still be in the process of up-grading (their F-5G has recently been redesignated F-20). Every aspect of aviation tends to be evolutionary rather than revolutionary: Northrop had identified a market need and filled it; the obvious next step was to consider a successor.

This meant staying in the export market rather than trying to produce a fighter for US service, and involved special considerations, the main one being technical and manufacturing participation by the main customer countries. By allowing the customer to have a hand in producing the aeroplane, the deal could be made much more attractive, as much of the money spent would benefit indigenous industry. The overall cost had to be attractive, both in capital outlay and in terms of operational life expenditure. The aeroplane would have to be mission-capable; while it could hardly approach the standards of the superfighters then under development, it would have to do a convincing job for the money. Cost-effectiveness was not a term current at the time, but this was the aim. In the event, the success of the cheap but potent F-5 has probably proved a hindrance to early sales of the new fighter. Finally, any

Right: This unusual planform view of the prototype Northrop YF-17 high over the Mojave desert shows many of the unique features of this fighter. The cut-outs in the LEX, through which the ground can be seen, the sharply swept, high aspect ratio horizontal tailplanes and the forward set of the vertical tail surfaces are all clearly apparent.

deals with foreign countries had to be politically acceptable to the DoD; the United States was hardly likely to countenance sales to a friendly country which was likely to use the equipment in a conflict with another friendly country.

Left: An early picture of the YF-17. The small strakes on the nose cone have not yet been added.

In the mid-1960s, Northrop initiated discussions with F-5 users and other interested parties to determine their requirements. They also postulated the most likely threat, and as many of the early F-5 operators were members of NATO, the discussions took on a distinctly European aspect. The fighter threat at that time was of course the Soviet-designed MiG-21, which was in service in large numbers with the Warsaw Pact countries: the projected threat was its successor, expected to be a simple, cheap, lightweight air superiority fighter, likely to enter service in the mid-1970s.

From the requirement studies emerged five basic missions: interception, air superiority, reconnaissance, close support and interdiction. These roles have conflicting requirements and an aircraft optimized for any one of them would be compromised in the others to a greater or lesser degree.

Lee Begin takes charge

Back in 1956, Northrop's Lee Begin, Jr., had made the first drawing of the N-156, later to become the F-5. Now, ten years later, he took charge of the Northrop project office team whose function it would be to translate ideas and requirements for the new fighter into hardware. Early studies showed that optimization for air superiority would result in the minimum compromise for the other roles. This was hardly surprising, as the aerodynamic requirements of an air superiority fighter themselves demand a great deal of compromise.

The role requirements are: (1) high rate of climb; (2) fast acceleration; (3) high turn rate combined with small turn radius; and (4) good transient performance (i.e., the ability to change the direction of flight rapidly). Items (1) and (2) require a high thrust/weight ratio, plenty of specific excess power (P_s) and the lowest possible drag, all of which are achieved by wrapping the smallest possible body around the largest possible engine. Item (3) demands a low wing loading which in turn calls for a relatively large wing (with its attendant weight and drag), plenty of P_s, and a high aspect ratio. Item (4) is conferred by good performance in pitch, and particularly in the rolling plane, with a fast rate of roll and rapid roll rate acceleration. This calls for a relatively small wing with a low aspect ratio. The art is to produce the best compromise between these conflicting requirements.

By 1967 the decision had been taken to design a Mach 2 air superiority fighter with secondary capability in other roles. The potential market was assessed at about 3,000 aircraft, of which the Northrop share could be about one-third, and the main types the new fighter could be expected to replace were the F-5, the Lockheed F-104 Starfighter and the Dassault Mirage III.

Gradually the outlines of what was to become the P-530 Cobra emerged. The original 'paper aeroplane' formulated in 1966 clearly showed its F-5 ancestry. The wing was of similar shape although greater in area, and featured a small leading edge extension (LEX) at the root, although it was high mounted. The engine inlets were set forward, just behind the cockpit, while the vertical and horizontal tail surfaces were almost identical. By the following year, larger LEX had been added, and the inlets were now positioned beneath the wings, which had developed a taper on the trailing edge from a point at about one-third of the span to the fuselage.

Above left: The Northrop F-20 Tigershark is, like the Hornet, descended from the F-5A, which the F-20 closely resembles.

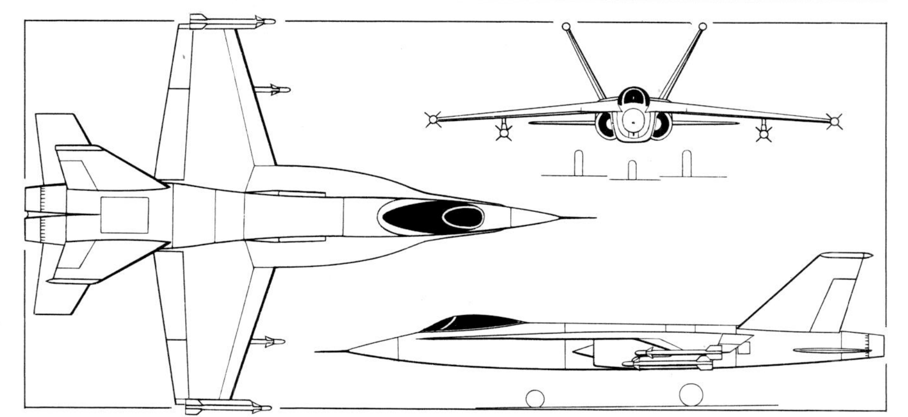

Left: Development stage – the P-530 Cobra. The early requirement for Mach 2 capability is reflected in the Starfighter-like fuselage shape and the half cones to the inlets.

Below: F-5E of the 425th TFTS. The success of Northrop's earlier lightweight fighter has probably inhibited sales of the company's F-18L.

YF-17 configuration: seven stages of development

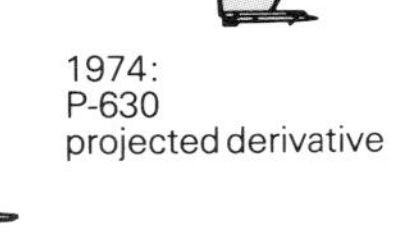

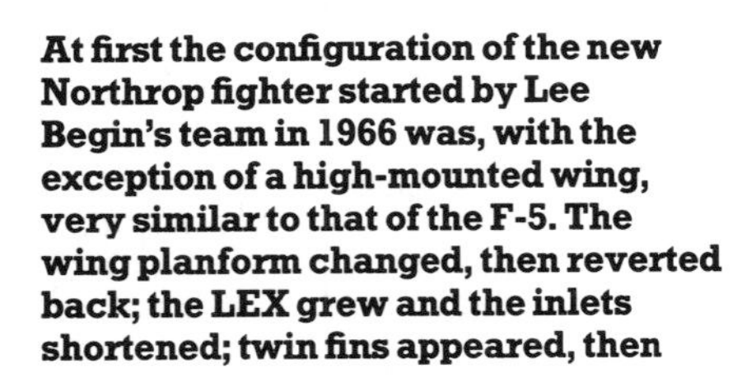

At first the configuration of the new Northrop fighter started by Lee Begin's team in 1966 was, with the exception of a high-mounted wing, very similar to that of the F-5. The wing planform changed, then reverted back; the LEX grew and the inlets shortened; twin fins appeared, then were moved forward, while the horizontal tail altered shape; and gradually the various features were refined. Aircraft design is evolutionary rather than revolutionary, and this illustration clearly shows how the YF-17 grew from the original F-5.

The most radical departures in layout came in 1968. The wing had reverted almost to its original shape, while the LEX had been extended much farther forward. They were now definite strakes rather than mere extensions, and their function was to generate large vortices across the upper surface of the wing which would inhibit the spanwise movement of the boundary layer air across the wing surface to give better handling qualities at high angles of attack (AOA). They also provided a destabilizing effect at transonic and supersonic speeds as the wing centre of lift moved aft, thus reducing trim drag. Finally the LEX served as compression wedges, reducing the intake Mach number by lowering the local AOA at the intakes. Running along the inside edge of the LEX were cut-outs, or slots. The purpose of these was to draw off the stagnant fuselage boundary layer air before it could be ingested by the engines, and expel it into the low pressure area above the wing roots.

Tail redesign

The tail had undergone the most radical changes. The horizontal surfaces were more sharply swept than before, and were of increased area, but the single fin and rudder had vanished, to be replaced by small twin vertical surfaces canted outwards, set above the engines at the extreme rear of the fuselage. Wind tunnel tests proved this layout unsatisfactory, and in 1969 the design showed that the LEX had been contoured, and the vertical tail surfaces were much larger, and had been moved forwards on the fuselage to a position where they overlapped both the trailing edge of the wing and the leading edge of the horizontal tail. This position produced some area rule effect, and good lateral stability was provided by the vortices from the LEX impinging on the fins, which were canted outwards at the startling angle of 30deg to obtain maximum benefit.

To achieve a thrust/weight ratio close to or exceeding unity, the two engines were required to have a static thrust in afterburner of about 12,000lb (5,445kg) each. The two main contenders for the supply of the engines were Rolls-Royce, with the RB.199, and General Electric with their new GE 15. General Electric agreed to develop their engine specifically for the Cobra, and the GE 15, later to become the YJ101-GE-100 was chosen. The twin-engined configuration selected was a follow-on from the F-5, giving extra safety and reducing operating costs due to attrition. Much controversy surrounds the twin versus single engine debate; it is believed, although figures differ widely, that a twin-engined design gives an engine-related attrition rate about 60 per cent of that suffered by single-engined designs.

By 1970 the P-530 Cobra looked much as the F-18 does today. The engine inlets had been shortened and the vertical tail surfaces were not canted at quite such an extreme angle. Meanwhile Northrop had set up other design teams. One team produced a layout for a single engined variant, the P-610, while another team, for reasons best known to the manufacturers, came up with a design called the P-600, almost identical to the P-530.

Unlike many fighters designed earlier, the Cobra always featured guns as part of its armament. At one stage two 20mm M39 revolver cannon were featured, but the final choice was the six-barrel General Electric M61, mounted under the nose. Wingtip launch rails were provided for Sidewinders, although theoretically the choice of missile was left to the customer, and seven hardpoints were provided for the carriage of up to 16,000lb (7,260kg) of stores. A one-piece bubble canopy gave the pilot excellent all-round vision.

By 1972 Northrop had invested a great deal of money in the Cobra project. More than 4,000 hours of wind tunnel testing had been completed and nearly 750,000 engineering man hours expended. Now they needed partners to further the programme who would be prepared to invest $100 million in the construction and testing of two pre-production machines, and whose requirements would total between 300 and 400 Cobras. The pre-production machines would allow the customer to evaluate the aeroplane before placing a firm order, and the initial payment would be offset against the total cost of the order. Provision had been made to allow the Cobra to be split up into production packages to share manufacture between customer nations, and both Holland and Norway were showing interest, but the vast capital expenditure involved in replacing a major portion of a nation's air force makes it a political and financial as well as a military issue. In practice, politicians become both military and financial experts and the resulting hot air delays the proceedings.

Hard lessons

Meanwhile wars had been fought and lessons learned. Vietnam 1965-1972, the Middle East wars of 1967 and 1969-70 and the Indo-Pakistan conflict of 1971 had all involved much air fighting, and one thing had become evident: close combat, the old-fashioned dogfight, was still very much a part of air warfare. The F-4 Phantom, arguably the most capable aircraft of its era, had found itself hard pressed by the comparatively cheap, lightweight Soviet designs. One of the truisms of air combat is "Don't fight the way your opponent fights best", but the Phantom often had no choice.

The historical record had until that time shown that the fastest fighter possessed a clear advantage in combat, both in attack and in defence. Until the Korean War, the difference between combat cruising speed and maximum speed rarely exceeded 15 to 20 per cent, but with the advent of Mach 2 capable fighters the difference became a factor of between two and three. Consequently, the advantage passed from the fastest aircraft in terms of capability to the fastest moving aircraft at the time of engagement. Over Vietnam, this was often the defending MiGs. The attempt to find an answer took two forms: better weapon systems with reliable beyond-visual-range (BVR) kill capability, and fighters which could beat the MiGs at their own game.

Two men who played a large part in developing the latter concept were defence systems analyst Pierre M. Sprey, and USAF Major John Boyd. Major Boyd is well known for formulating

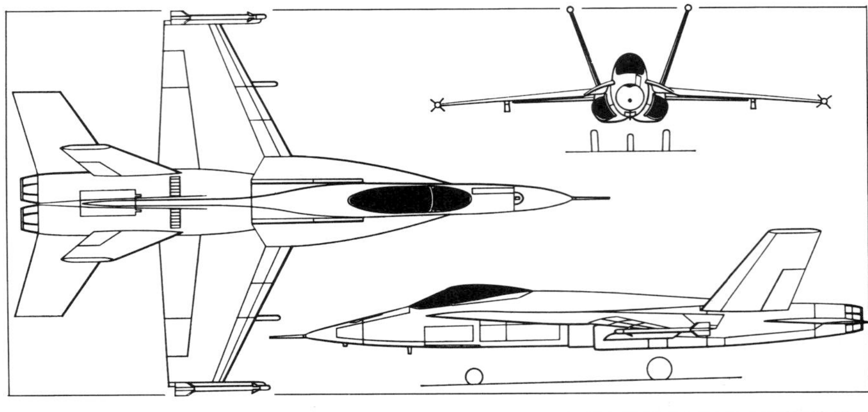

Right: The use of ultra-violet light allows the photography of airflow patterns formed in the wind tunnel. This picture of a YF-17 model on test, taken nearly a year before the first flight, shows the vortices formed by the LEX leading edge extensions.

the concept of energy manoeuvrability, and perhaps less so for his development of a flying technique to counter roll reversal by using rudder instead of aileron, 'setting the hook' as he called it. Their early studies centred around a design for a dedicated air superiority fighter, and priorities were established from the lessons of the past. The most obvious lesson was that the majority of air victories were the result of a surprise attack, so the problem became a question of how best to achieve surprise. Part of the answer lay in keeping the fighter small and therefore difficult to see. This was at the time almost heresy. The latest US fighters were the Grumman F-14 Tomcat and the McDonnell Douglas F-15 Eagle, both very capable, colossally expensive and extremely large.

Doubts were beginning to surface as to whether the established trend was the correct one. There could be little doubt that however capable a fighter was, it would stand little chance in close combat if heavily outnumbered by simple, austere (and cheap) fighters. This was in fact borne out by later Red Flag exercises, where the big super-fighters performed wonderfully well in numerically small engagements but achieved a kill ratio in multi-bogey combats barely in excess of 1:1 against F-5Es of the Aggressor Squadrons.

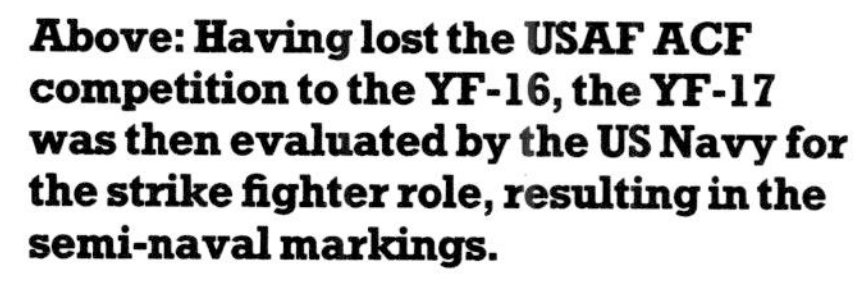
Above: Having lost the USAF ACF competition to the YF-16, the YF-17 was then evaluated by the US Navy for the strike fighter role, resulting in the semi-naval markings.

Below: Trailing smoke, a YF-17 lifts off the runway. Compare the flimsy landing gear and single nose wheel with the beefy-looking undercarriage subsequently added to the F-18.

Below left: Three-view drawing of the YF-17, showing the basic configuration finalized in 1973.

Below: The P-600 differed from the YF-17 in wing and tailplane planform, and in the cant angle of the fins.

Below right: The two YF-17s were later redesignated F-18L by Northrop for evaluation purposes.

Bottom right: The hooded aspect of the YF-17, which inspired the name Cobra, is seen at takeoff.

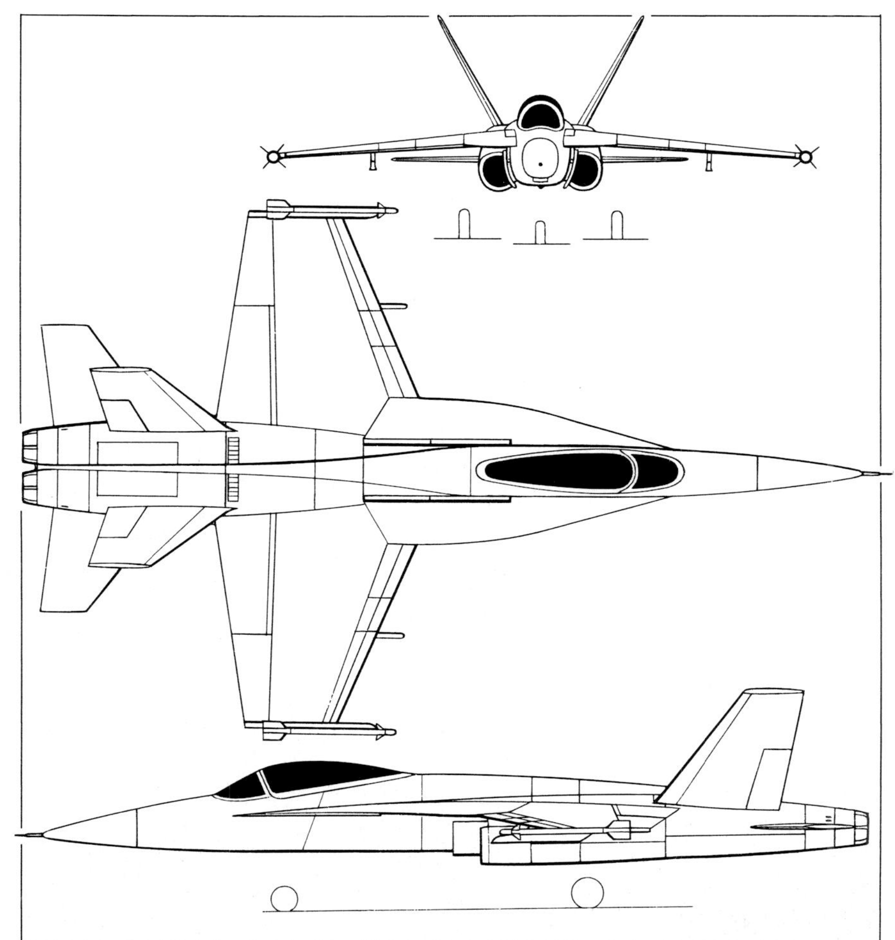

By 1971, Major Boyd was working for the Air Force Prototype Study Group, and was in a position to push the light fighter concept. At the same time, the Department of Defense resorted to the time-honoured custom of ordering prototypes which could be evaluated by flying against each other. Prototyping was regarded as a systems management technique to be pursued as a possible method of achieving continuing technical superiority cheaply, or as the jargon has it, 'in an austere funding environment'. It can be summarized as "let's see the goods before we buy".

LWF requirement

On January 6, 1972, the USAF issued a request for proposals for a lightweight fighter (LWF). Little in the way of performance minima was specified, thus freeing the designer to concentrate on the main requirements, which were to demonstrate exceptional manoeuvre and handling capability in the transonic regime. In short, the LWF was to be designed for greatest effectiveness in the middle of the flight performance envelope.

A minimum load factor of 6.5g was specified, along with limited avionics for navigation, communications and fire control for guns and missiles. The new design had to demonstrate advanced technology while keeping both weight and cost down.

LWF proposals were submitted by Boeing, Ling-Temco-Vought, General Dynamics and Northrop. The Northrop proposal was based not on the P-530 but on the virtually identical P-600. On April 13, 1972, the field was narrowed to two, contracts being placed with General Dynamics and Northrop, worth $38 million and $39 million respectively. Each was to build two prototype fighters for evaluation. They were to be technology demonstrators, and no Air Force requirement was to be presumed. A cost limit of $3 million was set based on a procurement of 300 aircraft in Fiscal Year 1970 terms. This allowed the competing design teams to make cost/performance tradeoffs, rather than be forced to keep tweaking the performance up a little to meet firm specification requirements, usually at disproportionate cost.

Bearing the Air Force designation YF-17, the first Northrop prototype was rolled out on April 4, 1974, its maiden flight took place on June 9, and it was followed into the air by the second prototype on August 21 of the same year. It had taken nearly eight long years, but the concept had finally made the transition from paper aeroplane to flyable hardware. The YF-17 was a single-seat fighter powered by two afterburning General Electric YJ-101 low bypass ratio turbojets, each rated at about 15,000lb (6,800kg) static thrust. The wings were set in the mid-fuselage position with 5deg of anhedral, and were of trapezoidal planform reminiscent of the F-5, with an area of 350sq ft (32.52m^2), a 20deg sweep at quarter-chord, with LEX and slots as drawn for on the P-530. Variable camber was featured, in the form of leading edge manoeuvring flaps and plain trailing edge flaps which deflected automatically as a function of AOA and Mach number.

The fuselage had grown to about 4ft (1.22m) longer than the P-530, and narrow strakes had appeared on both sides of the nosecone. The pilot sat on a Stencel Aero 3C ejection seat, which was raked back at an angle of 18deg. The bubble canopy gave excellent rearward visibility although this was somewhat negated by the airbrake when extended, this being located on top of the fuselage between the twin fins, which were canted outwards at about 20deg, rather less than those of the P-530. The all-moving tailplane showed most changes, being mounted low on the rear

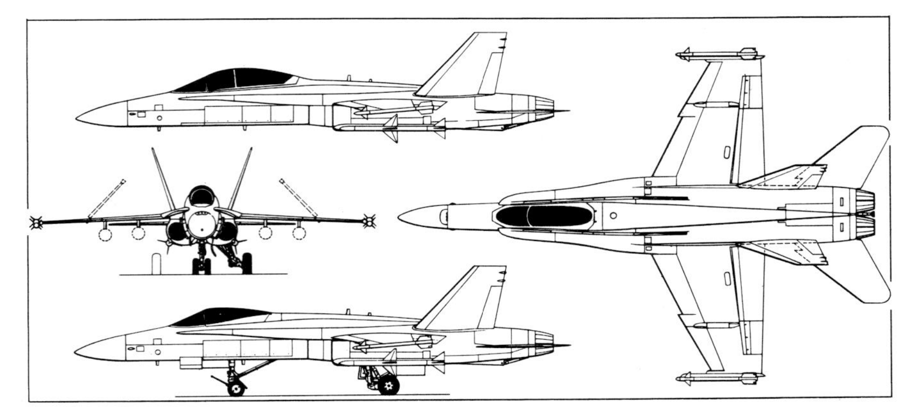

Above: The first prototype F-18 on an early flight demonstrates its variable camber wing.

Top: Hornet 1 looks rather weary after nearly four years of test flying. Navy and Marine Corps markings are on opposite sides.

Right: Hornet 3, the carrier suitability trials aircraft, flies over USS *America* on October 30, 1979, prior to making the first deck landing. During the next four days Hornet 3 carried out what the US Navy called, "the most successful sea trials in naval history", including 32 catapult launches.

Left: The original configuration of the F-18, plus a profile of the two-seat TF-18. The notched wing and tailplane leading edges are McDonnell Douglas innovations, along with the arrester hook and folding wings needed for carrier operations, and the altered tailplane planform.

Below: A contrast in prototypes. The YF-17 (left) is noticeably smaller and less chunky than the Hornet 3, and while both aircraft carry Sidewinders on wingtip rails, the F-18 also carries two Sparrows under the fuselage.

fuselage and swept more sharply than on the Cobra, while its span had increased to 22.21ft (6.77m) and its aspect ratio had risen considerably. Fly-by-wire (FBW) systems operated ailerons, rudder and tailplanes, with mechanical pitch and roll backup for the tailplane.

The YF-17 retained the nine external store stations of its predecessor and its air-to-air armament consisted of a Sidewinder mounted on a rail on each wingtip, and the M61 cannon mounted high in the nose instead of underneath as formerly. Avionics were basic: an air-to-air ranging radar by Rockwell International with a small phased array antenna; the Litton Industries LN-33 Inertial Navigation System (INS); a Teledyne transponder; and a gunsight head-up display (HUD) by JLM International. The clean takeoff weight had been held down to 23,000lb (10,430kg), which gave a very favourable thrust/weight ratio. Considerable weight savings had been achieved by redesigning the undercarriage, which on the Cobra had been intended for rough field operations and was consequently more rugged and heavier than necessary for runway operations.

In the air, the YF-17 performed well. During flight tests it demonstrated a top speed of Mach 1.95 (there was no requirement for Mach 2 and considerable weight and complexity savings had been achieved by using fixed inlets), a peak load factor of 9.4g, a maximum altitude of 50,000ft (15,250m), and a sea level rate of climb exceeding 50,000ft/min (254m/sec). Handling was excellent: the YF-17 could achieve AOA of up to 34deg in level flight, and 63deg AOA was reached in a 60deg zoom climb, while the aircraft remained controllable at indicated speeds right down to 20kt (37km/h). Northrop were consequently able to claim that their contender had no AOA limitations, no control limitations, and no departure tendencies. It was certainly an impressive performance.

A decision was taken in April 1974 that the LWF was no longer to be just a technology demonstrator, but that the successful contender would be developed into a USAF Air Combat Fighter (ACF). The general reasoning was that basic commitments demanded more fighters than the number of F-15s or F-14s that could be purchased with the funds available, so a nucleus of expensive high-technology fighters was to be supported by austere and much cheaper ones. This became known as the hi/lo mix, in which the fighter force was to have adequate numerical strength containing a significant level of high technology.

The flight test programme for the ACF contenders was rushed through in a few months instead of the normal two years. Formal evaluation took place towards the end of 1974, and the result was announced on January 13, 1975: the new air combat fighter for the USAF would be developed from the General Dynamics YF-16. The contest had been far from a walkover and the YF-17 had proved superior in some regimes, but the award had gone to its single-engined rival.

Navy fighter requirement

Meanwhile another potential market had emerged. Back in 1971 the US Navy had become concerned at the cost of the F-14 Tomcat, which had caused both rate of procurement and total number to be restricted to the extent that the Navy could not afford the number of Tomcats that it deemed necessary. Furthermore, the ageing Phantoms and Corsairs would need to be replaced in the not too distant future, and various alternative solutions were examined, including a cheaper F-14, a navalized F-15 and improved F-4s. A group called Fighter Study IV discussed and rejected these alternatives, formulating instead the requirements of a new Naval Air Combat Fighter with a secondary attack capability, known as VFAX (fighter/attack experimental aeroplane) and to be armed with both the short range Sidewinder and the medium range Sparrow. It was anticipated that this would be a totally new design, but Congress decreed that the USN should study derivatives of the ACF contenders.

Below: Maiden flight. Carrying dummy Sparrows and Sidewinders, Hornet 1 lifts off the runway at Saint Louis on November 18, 1978, with McDonnell Douglas Chief Test Pilot Jack Krings at the controls.

Northrop, inexperienced in the particularly demanding requirements of carrier-based fighter design, teamed up with McDonnell Douglas, makers of the very successful Phantom, and the result of this collaboration was a navalized version of the YF-17, known initially as the Northrop P-630 and later as the McDonnell Douglas Model 267. In retrospect, it seems probable that Congress had in mind the use of a common type by both Air Force and Navy, with cost savings resulting from a large order. The precedent had been set years earlier by the F-4 Phantom, which was designed for carrier operations and went on to form a major part of the USAF inventory.

Navy assessment

The US Navy took a long, hard look at both the ACF contenders. Quite apart from the constraints imposed by carrier operations, their requirements differed considerably from those of the Air Force. The F-16 was a close combat dog-fighter *par excellence*, whereas the Navy needed a machine that could be used to modernize both fighter and attack squadrons. In particular, fitting out a fighter to carry the medium-range Sparrow was a major task. The procurement limit set for the Tomcat was sufficient to equip only 18 of the 24 squadrons in the Carrier Air Wings, leaving a further six to be modernized, plus 12 US Marine Corps fighter squadrons, as well as six Navy and Marine reserve squadrons to be re-equipped. In addition, there were 24 front-line and six reserve attack squadrons then flying the A-7 Corsair, which would need new aircraft from the early 1980s. Allowing an annual attrition rate of 4.5 per cent, the result was a total requirement of 800 aircraft.

To fulfil the dual roles, the Navy needed a fighter which could easily be adapted to carry the Sparrow, and which could equally easily be converted for the attack role. The original P-530 had been designed as a strike fighter, so there were few problems in that direction. But the Navy, unlike the Air Force, carried out most of their missions over the sea. Engine malfunction leading to the loss of the aeroplane was bad enough, but over the sea it was also likely to cause the loss of a hard-to-replace piece of software – the pilot. General Dynamics had teamed with LTV (Vought) to produce a navalized variant of the YF-16, but the two engines of the Northrop/McDonnell Douglas fighter offered better safety, and this was a major factor in influencing the Navy's decision, although the YF-17 was also able to demonstrate better carrier recovery performance than its rival. A further consideration was that the design had more multi-mission potential.

The choice of the future F/A-18 as the Navy's new fighter/attack aeroplane was announced on May 2, 1975. A frequent accusation against the Navy during the following years was to be that they had bought the loser in the ACF competition. Of course, they had, but they had also bought the aeroplane that they considered was better suited to their needs. A comparison of USN requirements with the projected figures for the new fighter, now redesignated F-18, shows how closely they matched (see table). The only serious shortfall in the figures was on minimum approach speed, but it was felt that this could be improved.

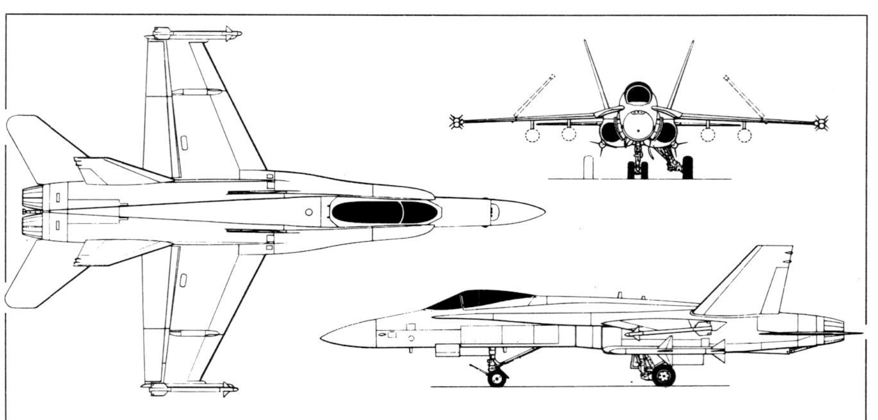

US Navy VFAX requirement/projected F-18 comparative data

	VFAX	F-18
Vmax dry power	Mach 0.98-1.0	Mach 0.99
Acceleration Mach 0.8-1.6	80-110 seconds	88.3 seconds
Combat ceiling (000ft/000m)	45-50/13.72-15.25	49.3/15.03
P_s, Mach 0.9/10,000ft (3,050m)	750-850ft/sec 229-259m/sec	756ft/sec 230.5m/sec
Buffet-free sustained load factor	5.0-5.5g	6.6g
tructural load factor	7.5g	7.5g
Single-engine climb rate	500ft (152m)/min	565ft (172m)/min
Minimum approach speed (kt/km/h)	115-125/212-230	131/241
Escort fighter radius (nm/km)	400-450/737-829	415/765
Strike radius (nm/km)	550/1,013	655/1,027

McDonnell Douglas, with the McDonnell Aircraft Corporation's extensive background of designing carrier fighters, became the main contractor for the F-18, with Northrop as a major subcontractor, the construction work being split approximately 60/40. At the same time, Northrop were to develop a land-based version of the fighter for the export market under the designation F-18L and, in the event of orders being placed, the work share was to be reversed. One thing was certain: the naval F-18 would be a much heavier beast than the YF-17, and more powerful engines would be required. This was resolved on November 21, 1975, when General Electric received a letter contract to proceed with the new F404, a developed and uprated version of the YJ-101. McDonnell Douglas received their letter contract on January 22, 1976, for the Full Scale Development (FSD) batch of 11 aircraft, nine single- and two twin-seaters: the first flight was scheduled for July 1978.

Adapting a land-based fighter to become carrier-capable is a very complicated process, and the aircraft also had to be fitted out to meet the Navy's dual-role mission requirements. Provision had to be made for the carriage of Sparrow missiles together with a compatible fire-control system, and all-weather avionics. A Hughes multi-mode radar was selected, and the nose had to be fattened by 4in (10cm) to accommodate the antenna, but no lengthening was needed.

Above: Dust billows up behind Hornet 3 as it carries out one of a series of cross-wind landing trials at Edwards AFB, California, during the summer of 1980. The flaps and ailerons are right down, and the stabilators are deflected to their maximum extent. A total of 119 landings were made in crosswinds of up to 30mph (48km/h).

Left: The Hornet after several fixes. The notches in the wing and tailplane leading edges have disappeared, as has almost all the slotted area in the wing root leading edge extensions.

Above: Hornet 3 aboard the aircraft carrier USS *Dwight D. Eisenhower* in February 1982. The complexity of the main landing gear, which retracts rearward while rotating through 90deg in order to avoid the Sparrow positions, is clearly apparent.

Below: The third FSD Hornet steps delicately (or so it seems) from the edge of the flight deck. As the weight comes off the wheels, the main gear takes on a totally different appearance (compare this view with the picture above).

To meet the long-range patrol requirements, provision for an extra 4,460lb (2,023kg) of fuel had to be made, bringing the total internal fuel capacity to 10,680lb (4,844kg) – compared with the 6,400lb (2,903kg) of the YF-17 – in four fuselage tanks plus an extra tank in the inboard section of each wing. A further 2,000lb (907kg) of fuel could be carried in drop tanks, and provision was made for in-flight refuelling.

For carrier operations a nosegear towbar and an arrester hook were added, and the undercarriage was redesigned to cope with the extra weight and the high stresses of catapult takeoffs and arrested landings, while to save space below or on deck, the outboard wing panels were made to fold. Alll these changes involved additional weight, bringing the projected gross weight of the F-18 to 33,580lb (15,232kg) at this stage, compared with the 23,000lb (10,433kg) of the YF-17. Considerable structural strengthening was required to cope with both the stress of catapult launching and arrested deck landings, or 'traps', and the weight increases. These modifications took the wing loading past acceptable limits. In consequence the wing area was increased from the 350sq ft (32.52m^2) of the YF-17 to 400sq ft (37.16m^2). This was done by increasing the span by 2.5ft (0.76m), and extending the chord by adding to the leading and trailing edges.

Much attention was paid to improving the carrier approach characteristics. The aerodynamic shape of the LEX was refined, and they were extended further forward on the fuselage, with a consequent increase in area. The deployment angles of the leading and trailing edge flaps were increased from 30deg to 45deg, and the ailerons were programmed to droop at a maximum angle of 45deg in low-speed flight. The horizontal tail surfaces, or stabilators, changed shape yet again, to give a lower aspect ratio than before, and a snag was added to the leading edges of both the wing and the stabilator to generate high energy air and reduce spanwise drift during carrier approaches.

Left: Stropped up on the catapult ready for launch. The rudders are toed in at a 25deg angle to provide a nose-up moment at takeoff.

It was predicted that these improvements would reduce the approach speed to 125kt (230km/h), at an AOA of 6-7deg, giving the pilot an excellent view over the nose. This compared very favourably with the Phantom, which approached nose-high at an AOA of 13-14deg. It was also anticipated that operation with very heavy payloads in zero wind over deck (WOD) conditions would be possible, and that the WOD requirements at maximum loads would be very low.

Cost and complexity

Another important factor that influenced final design, and one by no means unique to the Hornet, was the sheer complexity of modern fighters. This had a knock-on effect, as complexity caused costs to soar and greatly increased the lead time from design inception to service entry. Soaring costs also reduced procurement levels, increasing the temptation to soldier on with the existing equipment for a few more years, and this in turn increased the operational life of fighters already in service to unprecedented levels – the life span of the Phantom, for example, looks likely to exceed 30 years. As a result, the production Hornet was designed to have a very long service life of 6,000 flying hours, including 2,000 catapult launches and 2,000 traps. Low procurement levels also meant that fighters had to be designed to fulfil more than one role, which put survivability and maintainability at a premium. In time of war, a prime requirement is a high sortie rate. The Hornet was therefore designed for high survivability and extreme ease of maintenance, while its dual role was stressed by the unofficial but widely used F/A-18 designation.

At the same time, a single-seat aircraft was being asked to supplement the two-seat Tomcat, and replace both the single-seat Corsair and the two-seat Phantom. This was a hard act to follow, and the question is still frequently posed whether one man can handle the workload. McDonnell Douglas rose to the challenge and, in a *tour de force* of cockpit design, took the experience and technology gained on the F-15 and improved on it by a considerable margin. The hands on throttle and stick (HOTAS) concept was adopted, with all the controls necessary for the combat mission placed on either the throttle or the control column. This appears to have been remarkably successful, although it does seem to demand from the pilot some of the qualities of a concert pianist. Information is presented on three cathode ray tube (CRT) displays, thereby eliminating at a stroke most of the hordes of dials that cover the dash and side consoles of most modern fighters. The desired information is to be called up as needed by a little simple switchology.

First flight

The F-18 made its maiden flight on November 18, 1978, only four months late. Flown by McDonnell Douglas chief test pilot Jack Krings, the first Hornet, Bu. No. 160775, resplendent in a white, blue and gold paint scheme, lifted off the runway at Lambert Saint Louis International Airport at shortly after 1100 hours local time. The flight, during which no problems were encountered, lasted 50 minutes, and a speed of 300kt (550km/h) and an altitude of 24,000ft (7,300m) were recorded.

After the initial flight test phase at Saint Louis the first Hornet moved to the Naval Air Test Center (NATC), NAS Patuxent River, Maryland, for the FSD test programme, which was to last from January 1979 to October 1982. This was not to be without its tribulations but, after all, that is what test programmes are for. The eleven FSD aircraft and their functions are shown in the accompanying table.

Above: Hornet 6 in orange and white high-visibility livery. This aircraft was used for high AOA and spinning trials, for which the gaudy paint job was an asset.

Right: Different-coloured AIM-9s – port red, starboard white – assist aspect identification on film of the high AOA and spin tests.

Below right: Hornet 7 takes on fuel from a KA-3 tanker. The US Navy uses the probe and drogue system for in-flight refuelling rather than the USAF's boom and receptacle.

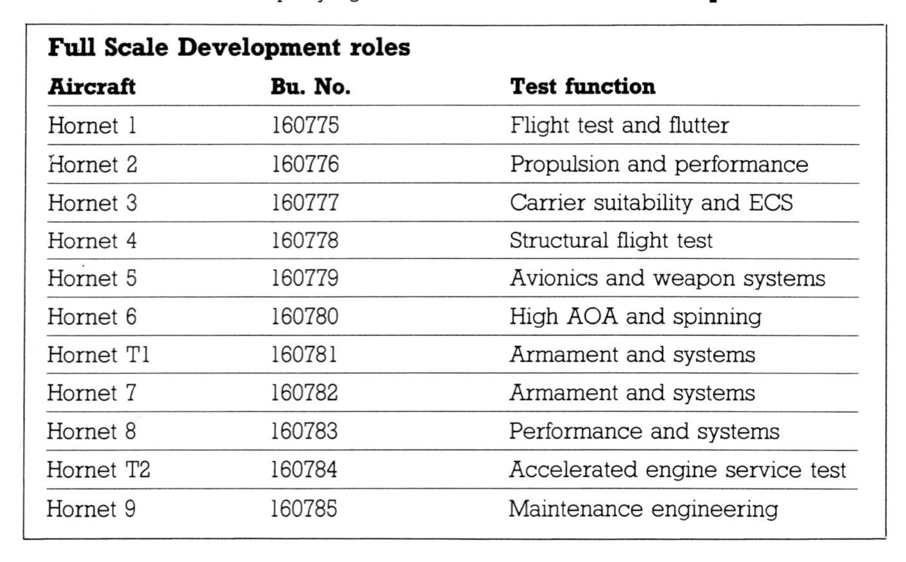

Full Scale Development roles

Aircraft	Bu. No.	Test function
Hornet 1	160775	Flight test and flutter
Hornet 2	160776	Propulsion and performance
Hornet 3	160777	Carrier suitability and ECS
Hornet 4	160778	Structural flight test
Hornet 5	160779	Avionics and weapon systems
Hornet 6	160780	High AOA and spinning
Hornet T1	160781	Armament and systems
Hornet 7	160782	Armament and systems
Hornet 8	160783	Performance and systems
Hornet T2	160784	Accelerated engine service test
Hornet 9	160785	Maintenance engineering

Previous test programmes had been carried out at a variety of locations, depending on which function was under test, but in the case of the F-18 the new Principal Site Concept was applied, with almost all flight testing taking place at Patuxent River. Apart from logistical advantages, this provided the opportunity for Navy and McDonnell Douglas personnel to work closely together, both in the air and on the ground, thus ensuring full naval participation in design improvements. The test programme called for no fewer than 3,257 test flights and involved two shifts, six days a week for both the Navy and the main contractor and the principal subcontractors. By mid-1981 over 3,500 flight hours had been logged in more than 2,600 flights.

The gestation period of the Hornet was surrounded by controversy. Some of this, mainly political in nature, was caused by the cost increases attendant upon the development of a lightweight fighter (a term thought to be synonymous with cheap) into a capable middleweight. Unfortunately the Hornet's development period coincided with a period of high inflation and this had the effect of making things look much worse than they were. The vast sums being spent had the effect of making the programme highly visible, and a considerable amount of speculation took place as to whether it would be cancelled.

With hindsight it appears unlikely that cancellation was ever a possibility, although alternative programmes were kept constantly under review. Potentially more damaging were the questions raised by the test programme itself. Certain technical problems and performance shortfalls emerged, some of which led to vociferous criticism, not only from the press and the politicians, but also from sections of the military whose personal prejudices ran counter to the F-18 concept.

An underlying cause was the proposed dual role of the new fighter. The F-4 Phantom, originally developed as a fleet defence fighter, had been turned into the greatest multi-role aircraft of its time. It had, however, been found wanting in the close combat arena, and there was a subsequent stress on the air superiority fighter role. Now here was McDonnell Douglas taking the *losing* fighter in the ACF competition and fitting it out to become not only a fighter but an attack aircraft as well. This gave some people the impression of a retrograde step, especially in view of the doubts that had been raised about the ability of one man to cope with the workload. Finally, there were those in the Navy who felt that the aircraft had been foisted on them by the DoD decision that the choice had to be made between the contenders in the Air Force competition, rather than commissioning a purpose built design. With this background, the Hornet had many detractors from the outset.

Early test flights confirmed that the nosewheel lift-off speed was unacceptably high, at 140kt (258km/h), and Hornet 3 (Bu. No. 160777) was modified to overcome this problem. The snag in the leading edges of the stabilators (a modification made by McDonnell Douglas) was

filled in, and a software programme change was made which automatically 'toed-in' the rudders 25deg at takeoff, while the weight was still on the wheels, providing a downward moment aft of the rotation axis. These modifications reduced the nosewheel lift-off speed to an acceptable 115kt (160km/h).

Hornet 2 (Bu. No. 160776) showed performance deficiencies which caused considerable acrimony. Navy Preliminary Evaluation (NPE) 1 revealed a shortfall of 12 per cent in specific range in cruise conditions, and a lesser but still significant shortfall in combat conditions, while acceleration from Mach 0.8 to Mach 1.6 at 35,000ft (10,670m) took longer than the 110 seconds predicted. A number of contributory causes were found. The engines were early standard, with performance levels below those subsequently achieved. The leading and trailing edge flaps were automatically actuated, and a programming fault had set the angle between two and three degrees too low to achieve optimum cruise performance. A software change cured this particular problem. There were also a number of faults in the environmental control system.

When all the changes had been made the shortfall on range reduced to 8 per cent, this being due to unanticipated drag. As a result, a Hornet was tufted to assess the airflow patterns on its surface. The slots in the LEX were found to cause a considerable proportion of the excess drag, and on Hornet 8 they were filled in for further tests. This did not cause the anticipated adverse effect on the airflow into the intakes and became a standard fix, along with two further drag reducing modifications: the wing leading edge radius was increased, and a fairing was installed over the environmental control system efflux under the fuselage to direct exhaust air rearward instead of straight out across the airstream. These fixes increased the range, but had little effect on the acceleration problem.

On the credit side, the engine response was excellent, the transition from flight idle to full afterburner taking less than four seconds, and afterburner lightup was satisfactory, being demonstrated at 45,000ft (13,700m), and 150kt (276km/h). Slight deficiencies in specific fuel consumption (sfc) were shown, but these were partially corrected by a change to the main fuel control.

Above: The dropping of objects from an aircraft in flight is a process fraught with peril: here an empty fuel tank is successfully discarded. For this series of trials in December 1980, Hornet 7 is equipped with a camera on the tailhook assembly and three more on each wingtip. In the event, the elliptical-section drop tank was replaced by a conventional tank.

Top: The seventh FSD Hornet built was the first two-seater, T1. Initially wearing white, blue and gold livery, it is seen here on October 26, 1982, in orange and white at NATC Patuxent River. Behind it are Hornet 6, also in high-visibility finish, and Hornet 3 with wings folded and Sidewinders still in position on the wingtips.

Below: Hornet T2 in US Navy low-visibility grey finish. Although the fuel capacity of the two-seater is reduced, it remains fully combat-capable.

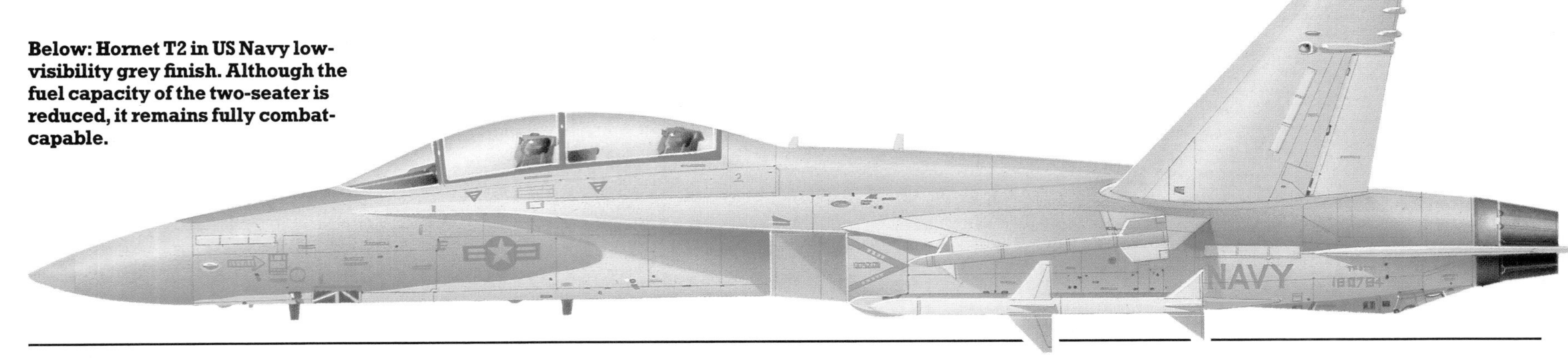

The range shortfall was to have serious repercussions. The Navy specification called for a range of 444nm (818km) in the fighter role and 635nm (1,170km) for attack missions. As at November 1979, the range limits had been set at 404nm (745km) in the fighter and 580nm (1,068km) in the attack configurations. Additional fuel tanks were not an acceptable solution, since the initial specified weight of 20,146lb (9,138kg) had already been exceeded by 1,962lb (890kg). An excess of 1,600lb (762kg) was considered acceptable, and a weight reduction programme was instituted to save 341lb (155kg), although this was not to take effect until the 123rd aircraft.

The most serious problem of all concerned rate of roll, which was well below the specified rate of 180deg/sec. Figures released early in 1980 gave the achieved roll rates as 185deg/sec at Mach 0.7, 160deg/sec at Mach 0.8, and 100deg/sec at Mach 0.9, all at 10,000ft (3,050m). At 20,000ft (6,100m) the roll rate was on specification at Mach 0.9, but as velocity increased so roll rate diminished. Analysis and observation showed the problem to have two main causes, namely flexing of the outer wing panels, and roll damping with the wingtip Sidewinders in place: when the outboard aileron was deflected in the transonic speed range, the wing bent in the opposite direction to counter the aileron action. The leading edge could actually be seen curling up from the cockpit!

The solution was a compound one. The snag in the leading edge, which incidentally did not feature in the YF-17, was eliminated, and the trailing edge box was strengthened to increase torsional stiffness, at the same time strengthening the trailing edge box into the wing root. This involved using monolithic graphite material instead of the sandwich in the inner wing and aluminium in the outer section. The trailing edge spar was also thickened, together with its webs and caps, while the ailerons were extended outboard to the wingtips, increasing their area by 36 per cent, and differential movement of the leading and trailing edge flaps was introduced. The differential horizontal authority of the stabilator was also increased. A spin-off effect of the increased aileron size was a 7kt (13km/h) reduction in the undesirably high carrier approach speed.

The Hornet encountered certain other problems during the test programme. The No. 4 fuel cell in the fuselage was very prone to leakage, and only a redesigned and strengthened cell cured the problem. Structural testing also revealed some flaws, details of which are given in the following chapter.

Carrier suitability trials

Hornet 3 (Bu. No. 160777) was slated for carrier suitability trials. After extensive testing at Patuxent River, during which over 70 catapult launches and 120 arrested landings were made, Navy test pilots Lt. Cdr. Richards and Lt. Grubb flew out to the USS *America* for initial sea trials during the late afternoon of October 30, 1979. During the next four days the Hornet carried out what were later described as "the most successful sea trials in Naval Aviation history". Between them, the two pilots carried out 32 catapult launches and traps, plus 17 touch-and-go landings, or 'bolters', and demonstrated vertical descent rates of 19.5ft/sec (5.9m/sec). Serviceability was 100 per cent throughout the trials, with no hold-ups recorded, and general characteristics, including deck handling, were recorded as excellent. Most catapult launches were made using intermediate power although full afterburner was used during two. The on-board auxiliary power unit (APU) was used for starting the engines, thereby keeping the 'yellow stuff' on deck to a minimum.

However, on its return from the trials Hornet 3 blotted its copybook when, arriving at NAS Oceana, it suffered, of all things, a landing gear failure. The fault lay in the centering mechanism to the main gear axle, which was experiencing a higher than predicted stress level. A new dual-chamber shock strut had to be developed, and this cured the fault.

This same aircraft was detached to Edwards AFB in California during the summer of 1980 for crosswind landing tests, and a total of 119 landings were made in crosswinds up to 30mph (48km/h). On March 17, 1981, Hornet 3 was again in trouble when, after a high sink rate approach at Patuxent River, the end of the fuselage-mounted rod on one of the main gears pulled out of the side

Below: In the attack role, FLIR and LST/SCAM pods replace the Sparrows on the fuselage mountings.

Above: Hornet 8, assigned to performance and systems tests, poses for the camera during IFR trials in May 1981. The refuelling probe can be seen extended, and the tanker is, slightly unusually, a USAF KC-10 with drogue gear.

Left: The ninth single-seat Hornet to be built was the first to receive the low-visibility grey finish. Seen here in fighter configuration, it is unusual in having no FSD number on the tailfin.

Below: With a payload of four Mk 84 1,000lb (454kg) slicks, this aircraft – also shown opposite – demonstrated the Hornet's attack range with the simulated attack on the Pinecastle range, 620nm (1,150km) from NATC Patuxent River, in September 1981.

brace actuator. After touchdown, the gear moved outboard and collapsed, causing damage to the flaps, wingtip, and engine intake.

The first fully automatic hands-off landing was made at Patuxent River on January 22, 1982, with McDonnell Douglas test pilot Peter Pilcher at the controls. This was the Hornet's first flight carrying the automatic landing system, and it is believed that this was the first time a fully automatic landing had been made on an initial flight. On January 26 Navy pilots started using the system, and by August of that year sufficient experience had been gained for the second batch of sea trials to be made, this time aboard the USS *Carl Vinson*. Two Navy pilots shared 63 catapult launches and traps, and numerous bolters, with various load combinations both by day and by night

Weapons trials

Other tests were going well, and the air-to-air and air-to-ground weapons trials were particularly successful. The first live missile firing was carried out by Hornet 5 (Bu. No. 160779) during December 1979, with McDonnell Douglas test pilot Bill Lowe at the controls, and the Sidewinder passed within 2.5ft (0.76m) of the BMQ-34 radio controlled target, well within lethal range. By October 1980 eight missile firings with both Sidewinder and Sparrow had been carried out against radio-controlled drone targets, with a 100 per cent success rate. No fewer than five of the eight missiles scored direct hits.

The ground and air firing tests of the M61 cannon also proved satisfactory. The ground tests involved firing the complete 570-round magazine in one long burst, while in the air tests six short bursts were used to empty the magazine. Both the 4,000 and 6,000 rds/min modes were used, and firing the gun was found to have no detrimental effect on either radar tracking or engine operation. The gun position above the nose was suspect, since it was thought that it might interfere with air to ground visual tracking at night, and that the accumulation of gas particles on the windscreen might degrade night visibility, so another test involved firing at a flare on a one-man liferaft on a cloudy night. No tracking difficulty was encountered, and no problems were caused by the gas particles.

Nor was the attack capability of the Hornet neglected. After early release trials of various stores, following which it was found necessary to move the store racks 5in (12.5cm) forward due to a flutter problem, one of the batch of nine pilot production Hornets (Bu. No. 161248) took off from Patuxent River bound for the Pinecastle Range Complex near Orlando, Florida, some 620nm (1,150km) away. It carried four Mk 83 1,000lb (450kg) bombs, two AIM-9 Sidewinders and three 315US gall (1,192lit) external tanks, with a Martin-Marietta laser spot tracker and Perkin-Elmer strike camera (LST/SCAM) pod on the right inlet position, a Ford Aerospace forward looking infra-red (FLIR) pod on the left inlet and a full load of 570 20mm cannon shells in its magazine. The Hornet's gross takeoff weight was 48,253lb (21,900kg), and after depositing its ordnance on target it returned in just over three hours with 1,600lb (726kg) of fuel remaining.

One change that was found necessary involved the drop tank designed specifically for the Hornet. Manufactured from spun fibre impregnated with aluminium, it was elliptical in cross-section to give better ground clearance, but the stresses imposed by catapult launches and traps proved too much, and it was replaced by a cylindrical tank holding an extra 15US gall (57lit) of fuel.

The General Electric F404 engine, despite being a relatively new design, was one of the success stories of the programme. As a previously untried engine it was subjected to an accelerated test programme – referred to as the 'Hornet Hustle' – involving Hornets T2 (Bu. No. 160784) and 9 (Bu. No. 160785). In just 55 flight days the two Hornets flew 116 missions, totalling just short of 150 flying hours, and on three occasions Hornet T2 achieved six flights a day, a remarkable performance. The F404 was found to have excellent throttle response, to be virtually stall-free, and to have little trouble relighting in flight.

Above: A swarm of Hornets over the Mojave desert, as Naval Air Test and Evaluation Squadrons VX-4 from Point Mugu and VX-5 from China Lake join forces. All are single-seaters and all carry centreline tanks, while the rearmost aircraft also has wing tanks. Some carry Sidewinders while others have bombs on the wing pylons.

Serious incidents

Only three serious incidents occurred. Trouble with a No. 4 bearing caused an engine shut-down in flight, but an improved bearing in production engines overcame the problem. An engine fire shortly after takeoff caused a mission to be aborted, and on examination three non-adjacent turbine blades were found to be fractured, although none had pierced the engine casing.

The most serious incident occurred on September 8, 1980, the day after the Farnborough Air Show, when Hornet T2 took off en route for Spain; the pilot was Jack Krings, by now the Director of Flight Operations for McDonnell Douglas, with Lt. Col. Gary Post of the USMC in the rear seat. At about 18,000ft (5,500m) on the climb-out there was a loud explosion, followed by a rapid temperature rise in the right turbine. The engine was immediately shut down and an attempt was made to reach the A&AEE airfield at Boscombe Down, but the throttle for the left engine appeared to be jammed, and control problems were experienced. Finally Krings and Post were forced to eject, at a speed of about 400kt (740km/h) and an altitude of 4,000ft (1,200m). The stricken Hornet crashed at Middle Wallop, Hampshire.

The affected engine was a pre-production model which contained a different type of material in the low-pressure turbine from that used in production standard engines. It had done about 300 flight hours at the time of the failure, which was due to the low-pressure turbine fracturing in flight. After extensive testing it was recommended that some redesign work be undertaken on certain rotor parts, although the fact that flight testing was resumed shortly afterward indicates that the failure was of a type unlikely to recur frequently.

The final FSD Hornet successfully completed its portion of the trials in October 1981. As well as the maintenance engineering inspection, it had participated in the Hornet Hustle and undergone electro-magnetic compatibility tests. The design case for the Hornet, based on the calculated radiation levels for the deck of an aircraft carrier, was 200V/m, compared with the 20V/m specified for the land-based F-16, and an average of 2V/m for the previous generation of aircraft. A couple of channels were found to have insufficient protection and new filter connectors were provided in the flight control system.

Climatic testing was carried out at the McKinley Climatic Laboratory, operated by the USAF's 3246th Test Wing at Eglin AFB, Florida. The Hornet was exposed to rigorous weather tests in a gigantic hangar, with temperatures fluctuating between −65deg F and +125deg F (−54/+52deg C). Winds of up to 100mph (160km/h) were simulated, and the Hornet was subjected to precipitation ranging from monsoon rains falling at 20in (50cm) per hour to blizzards.

Pilot reports on the Hornet had been unanimous in their praise of its handling qualities – it had reached higher AOA without control loss than even its F-16 rival, and it appeared virtually spin-proof. It was therefore extremely puzzling when a pilot production aircraft (Bu. No. 161215) crashed in Chesapeake Bay on November 14, 1980. Lt. C. T. Brannon, a naval test pilot attached to test and evaluation squadron VX-4, was on a routine handling flight when he lost control at about 20,000ft (6,100m). Failing to recover the aircraft, he ejected, and was fished out of the sea unharmed.

Accident investigation

Hornet 6 (Bu. No. 160780), painted orange and white for high visibility and with an anti-spin chute carried in a small box mounted above and between the engine nozzles, had carried out the high AOA and spin tests successfully. The Hornet had been found to be reluctant to spin at all, and when the control settings for inducing the spin were released it tended to fly out of it with no further ado. Points well outside the USN specifications had been reached without difficulty, including 78deg AOA coupled with 12deg of yaw; 74deg AOA coupled with 25deg of yaw; and, amazingly, 65deg AOA with a sideslip of −15deg to +30deg.

Tests were immediately instituted to duplicate the flight conditions leading to the accident, at first with no success. No fewer than 110 spin trial flights were

Left: Bu. No. 161248, first of the pilot production batch, and the Hornet that flew the Pinecastle strike demonstration, turns low over the water as if to line up for a deck landing, though as the hook is not down this is hardly likely.

Above: An atmospheric scene aboard USS *Constellation* as an aircraft launches at night, while three others of VX-5 stand by.

Right: A crowded deck on 'Connie' by day, with two Hornets on the catapults and a further four from VX-5 spotted, with wing tanks but no weapons.

made by both McDonnell Douglas and Navy pilots, and it took four weeks to duplicate Brannon's flight departure, which appears to indicate that it was an extremely unusual occurrence. On the first occasion, the pilot was able to regain control by cutting the engine on the outside of the spin to flight idle, while using full 'burner on the inside engine.

Once the problem had been identified, a solution was found. A spin recovery switch was added to the flight computer control which defeated the computer's logic and gave full control-surface authority to the pilot. In the test aircraft the cockpit displays were programmed to go blank when a yaw rate exceeding 15deg/sec was sensed, and the words 'spin recovery' appeared on the displays with an arrow indicating the direction the control column should be moved for recovery: once yaw reduced to less than 15deg/sec the pilot centred the controls and the aircraft flew out of the spin. With the switch added, Brannon's flight was duplicated for the second time, and the switch performed as advertised. The switch was incorporated on production Hornets, while improvements to the computer logic were put in hand to render it unnecessary on later models.

The training squadron VFA-125 was commissioned at NAS Lemoore, California, on November 18, 1980, and on February 19, 1981, it took delivery of its first Hornet. The way had been long, beset with difficulties, frustrations, and talk of cancellation. The Hornet had been roundly condemned as a can of worms, not as good as the types it was to replace, far too expensive and inferior to the contemporary F-14, F-15, and F-16. It is, however, worth noting that the criticisms came from those who had not flown it. It is a compromise aeroplane certainly, but a first class compromise.

Right: The sign outside MCAS El Toro, California, where the first operational Hornet unit, VMFA-314 Black Knights, converted from F-4Ns in January 1983. The loads shown – LGBs and targeting pods, Sparrows and Sidewinders – symbolize the dual role.

Structure

A modern fighter is a series of compromises, and trade-offs have to be made between many conflicting requirements. The transition from lightweight air combat fighter to middleweight multi-role carrier fighter and attack aircraft was not an easy one, with additional weight and the demands of carrier operations having to be accommodated without degrading performance. Nevertheless, with reliability, maintainability and survivability as fundamental design criteria, the F/A-18 has demonstrated an impressively high rate of availability, and the Hornet should be airborne to meet any threat as it arises.

Above: An unusual close-up belly view of FSD Hornet 7 taken during armament and systems trials in February 1980. The ECM fairings under the intakes and the chaff and flare dispensers show up well.

The Hornet is unusual in that there are two aircraft designs which externally appear almost identical, one known as the McDonnell Douglas F-18A Hornet, and the other as the Northrop F-18L Hornet. This unprecedented situation arose through the navalization of the Northrop YF-17 by the Saint Louis-based McDonnell Aircraft Company. MCAIR, as the company is known, is a division of the McDonnell Douglas Corporation, and was nominated as the main contractor for the F-18 due to its wide experience in the design and development of carrier fighters, with the Northrop Corporation as the major subcontractor, the work being split approximately 60/40.

Joint manufacture

McDonnell Douglas builds the forward fuselage and cockpit, wings, stabilators and landing and arresting gear, while Northrop manufactures the centre and aft fuselage sections, the splice between them, and the vertical fins. The joined fuselage sections, complete with all their associated plumbing and systems, are then shipped to Saint Louis for final assembly by McDonnell Douglas, who, with their carrier fighter background, retain overall responsibility for stress analysis. Major systems such as hydraulics, fuel, engines, environmental control, and secondary power are the responsibility of Northrop, while the crew station, avionics and flight control systems are down to McDonnell Douglas.

Both companies used full scale and very accurate engineering development jigs to work out the precise routings and positions of the plumbing and subsystems within the airframe. In the event of orders being placed for the F-18L, a lighter and simpler (and, some think, more potent) version of the F/A-18, the main contractor/subcontractor relationship would be reversed and Northrop would take the project leadership with a 60 per cent share of production.

The original Hornet contract was for eleven FSD aircraft, comprising nine single-seaters and two twin-seaters, plus one fatigue test and one static test airframes. Based on a purchase of 800 aircraft, the cost was predicted at $5.9 million each in Fiscal Year 1975 terms, including the cost of engines and avionics which were to be supplied through separate government contracts. Unfortunately, the cost was to escalate out of all proportion during the next few years, a period of very high inflation, and both politicians and the popular press were heard calling for cancellation.

In all fairness, the same vociferous protests had greeted just about every new American fighter project during the previous ten years, the F-111 being the outstanding example. A type of perverse logic seemed to prevail: if it was cheap it must automatically lack capability, whereas if it was capable it must be either too expensive or too complicated to work properly, and the slightest setback was blown up to assume the proportions of a major disaster. Fortunately, more sober counsels prevailed and a total of 1,377 Hornets, including the 11 FSD aircraft, were ordered for the US Navy and Marine Corps, although the USN stated that their actual requirement was for 1,845 aircraft, despite unit costs having risen to over $20 million in the interim. The rate of production had reached seven Hornets a month in December 1983 and is planned to peak at 18 per month in 1987, and to continue, albeit at a lower rate, until 1994.

Technically, the Hornet structure represents a transitional stage in fighter design. By weight, 49.6 per cent of the airframe is made up of aluminium, while steel accounts for 16.7 per cent, titanium 12.9 per cent and advanced composites 9.9 per cent. These proportions compare interestingly with the Air Force F-15 and F-16, which are at extremes of the scale. The F-15 contains by weight only 37.3 per cent aluminium and 5.5 per cent

Right: The structural flight test Hornet pictured in June 1979. Without paint the various materials are visible: the black areas are graphite/epoxy composites while the dark grey patch on the fin is titanium.

Below: A different view of the same aircraft. The composite areas under the LEX are access panels for the avionics LRUs.

steel, but 25.8 per cent of titanium, whereas the F-16, which deliberately avoided high technology materials, contains about 80 per cent of aluminium, just under 8 per cent steel, and only 1.5 per cent titanium and less than 3 per cent of advanced composites.

The use of advanced composites in the Hornet was, until the advent of the AV-8B Harrier II, more extensive than in any other operational fighter. Although contributing just under 10 per cent of the total weight, graphite epoxy composite material covers approximately 40 per cent of the surface area of the Hornet. Light in weight, of high strength, fatigue resistant and, most important in a carrier environment, corrosion resistant, it is used in the wing skins, trailing edge flaps, ailerons, stabilator, fin and rudder surfaces, most maintenance access doors, and the airbrake.

The fuselage is a semi-monocoque basic structure incorporating differential area ruling, with reduced area above the wings and increased cross-sectional area below them, both of which help to generate positive lift and reduce lift-induced drag. The forward position of the fins was also selected partially for its area-rule effect. The fuselage structure is mainly of light alloy, with machined aluminium fuselage frames. The engines are located in the extreme rear of the fuselage, with titanium firewalls between them. The engine bay face is formed by the production break between the rear and centre fuselage sections. The pressurized cockpit in the forward fuselage section is of fail-safe construction and is mounted above the nosewheel bay, which is probably longer in proportion to fuselage length than in any other aircraft. The 'barn door' type hydraulically actuated airbrake is mounted between the twin fins, and is of graphite epoxy material.

Two-dimensional inlets

Mach 2 was not a specified requirement, so the engine inlets are two-dimensional external compression types, allowing savings in weight and complexity. They also have a small radar cross-sectional area, thus reducing the risk of head-on detection. The inlets are preceded by 5deg fixed ramps, solid at the front and perforated just in front of the inlet proper in order to dispose of the sluggish boundary layer air from the ramp face. The only moving parts on the intakes are the bleed air doors, which exhaust upwards into the LEX flow field.

The wings are of cantilever construction, set in the mid-fuselage position with slight anhedral. Typically Northrop in their trapezoidal shape, they feature variable camber and, at first sight rather outlandish-looking, leading edge extensions (LEX). This combination, known as a hybrid wing, confers excellent manoeuvrability in the subsonic/transonic flight regime, and really outstanding high-AOA capability, rather better even than the vaunted F-16. The main wing construction is a six-spar machined aluminium alloy torsion box, with graphite-epoxy wing panels. The box is attached to the fuselage by six dual-fork attachment lugs.

Control surfaces

The variable camber is achieved with full-span leading edge flaps which have a maximum extension angle of 30deg, and single-slotted trailing edge flaps, actuated by Bertea hydraulic cylinders, with a maximum angle of 45deg. Computerized automatic actuation sets the optimum angle for the prevailing flight conditions, whether manoeuvre or cruise. The ailerons, with Hydraulic Research actuators, can also be drooped to an angle of 45deg, thus acting as full-span flaps to give low landing approach speeds, and the ailerons and flaps also provide differential movement for roll. The wing loading is modest and the variable camber gives the good gust response characteristics necessary for the attack role. The leading edge flaps and the ailerons have aluminium skinning, while the trailing edge flaps are of graphite epoxy. The wing fold essential for carrier stowage comes at the inboard end of each aileron, with a titanium hinge and an AiResearch .nechanical drive, and a Sidewinder launch rail is carried on each wingtip.

Above: Small is beautiful in the close combat arena, and this head-on shot demonstrates the Hornet's small presented area.

Below: The work split between McDonnell Aircraft Company, Saint Louis, and the Northrop Corporation in California is approximately 60/40. As shown in the diagram, Northrop's contribution (shown in red) comprises the rear fuselage and fins, which are shipped to Saint Louis for final assembly with the MCAIR-built forward fuselage, wings, stabilators and landing gear.

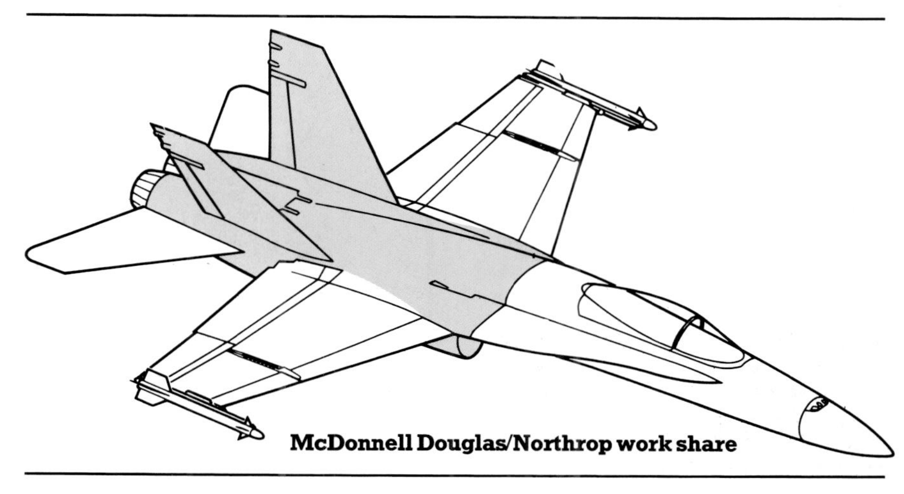

McDonnell Douglas/Northrop work share

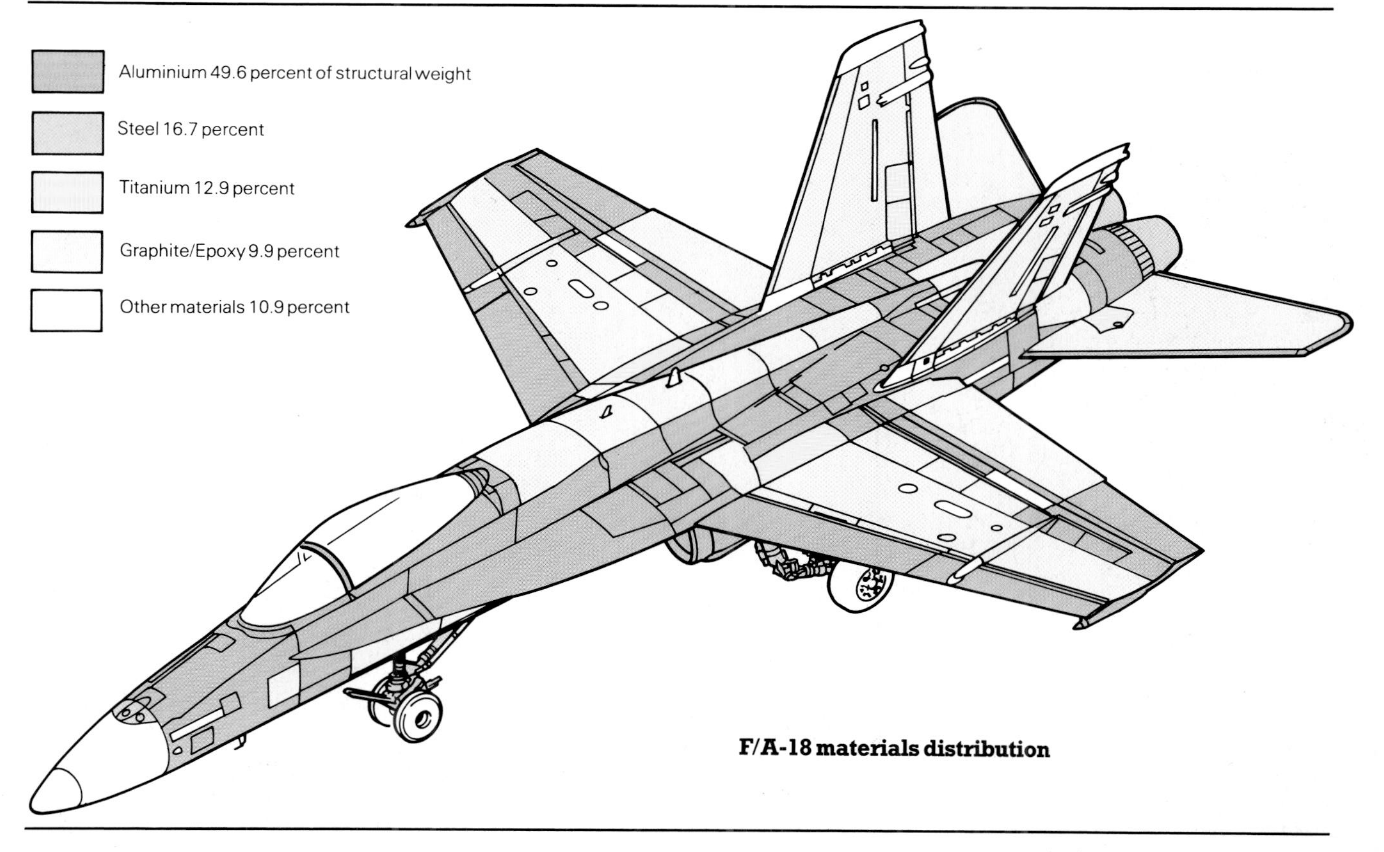

F/A-18 materials distribution

Right: Conventional materials and composites are combined in the Hornet's airframe for optimum strength and lightness. Graphite/epoxy composites cover 40 per cent of the surface area while accounting for under 10 per cent of structural weight. Titanium is used far more widely than in the F-16 but less than in the F-15.

Above: A VFA-125 Rough Raiders Hornet is prepared for a mission. The elliptical section drop tank being fitted gave greater deck clearance, but was unable to take the stresses of ship-based flight.

The most remarkable feature of the wing is the LEX, which extends forward past the cockpit. It acts as a giant vortex generator which scrubs the wing clean of slow-moving boundary layer air and permits controlled flight at AOA exceeding 90deg, although it should be noted that engine thrust is needed to maintain control of the aircraft in these regimes. The LEX also increases maximum lift by up to 50 per cent, and reduces lift-induced drag, supersonic trim drag and buffet intensity. Furthermore, it acts as a compression wedge to reduce the Mach number at the engine inlet face, and reduces the angle of the air entering the inlet by 50 per cent of the AOA.

Each LEX contains empty space that nothing other than fluid would fit into, but if the space were used for fuel it would be too far ahead of the centre of gravity, and damage would cause fuel spillage into the engine intake, which is undesirable to say the least. The LEX cannot house cannon without bulges spoiling the airflow, but one use has been found: the Hornet boarding ladder is integral, and retracts neatly into the underside of the left-hand LEX.

The Hornet's all-moving tailplanes, or stabilators, are unusual in that their span exceeds 50 per cent of the wing span, apparently to give adequate roll control as tailerons, where the effectiveness of the ailerons is insufficient. They are actuated by National Water Lift servo-cylinder hydraulic units, acting collectively for pitch and differentially for roll control. Like many modern fighters with engines mounted at the rear, the Hornet is close-coupled, and the pitch rates achieved have been described by Navy pilots as 'unbelievable'. The stabilators are constructed from aluminium honeycomb clad with graphite epoxy, but with aluminium leading and trailing edges, reinforced with titanium near the pivots.

Tail configuration

The twin fin and rudder arrangement has two outstanding advantages, in that it reduces or eliminates the effect of body vortices at high AOA, and also presents a smaller radar cross-sectional area than a single fin of the same total area seen from side-on.

The fins are of cantilever structure, with a six-spar torsion box connected to six fuselage/fin attachment frames with integral lugs. They are skinned with graphite epoxy, with titanium leading edges and detachable glass fibre tips. The mid-fuselage location and outward cant was selected to avoid blanketing by the fuselage at high AOA and also to avoid the possibility of biplane interference at low forward speeds. Titanium panels cover the rudder hydraulic actuator positions, and the rudders themselves are of one piece aluminium with graphite-epoxy skinning.

The retractable tricycle undercarriage is manufactured by Cleveland Pneumatic, and accounts for a great deal of the steel used in the Hornet, while the wheels and brakes, which are of the multi-disc type, are by Bendix. The nosegear consists of a forward retracting

Wing flap positions

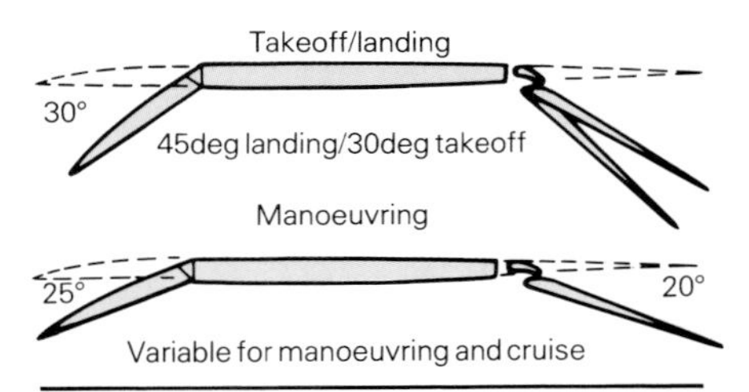

Above: The Hornet's wings feature variable camber, and computerized actuation which alters the section automatically to give optimum performance in all flight conditions. Variable camber also reduces gust response and improves the ride quality at low level and high speed.

Right: January 1982, and VFA-125 deploy to Yuma for ACM. A couple of adversary aircraft can be seen in the background, as can an AV-8A Harrier in hovering flight.

twin nosewheel, with a shuttle arm for catapult launching and a long drag strut extending rearward. Nosewheel steering is accomplished by an Ozone hydraulically actuated unit. The main gears have single wheels that retract rearward, turning through 90deg to stow horizontally in housings under the engine air ducts. The Sparrow missile positions are in the direct retraction path and must be avoided, resulting in heavier and more complex gear. The aircraft is designed for a maximum descent rate of 24ft/sec (7.32m/sec), a vertical speed of just over 16mph (25km/h). The tyres are made by B. F. Goodrich, and are size 22×6.6-10, 20-ply, on the nosegear, and 30×11.5-14.5, 24-ply, on the main gear. Pressures are 350psi (24.13 bars) all round for carrier operations, while for land-based operations the pressures are 150psi (10.34 bars) for the nosewheel and 200psi (13.79 bars) for the main gears.

The Hornet was the first production aeroplane to fly with a quadruple (four channel) digital fly-by-wire (FBW) control system. FBW is generally accepted as indicating pilot control of aircraft movement by means of electrical impulses, rather than pilot control of the position of the control surfaces. The pilot, using (in the Hornet at any rate) orthodox stick and rudder movements, signals his requirements to the flight control computer and lets the electrons worry about the vulgar details of flying the aeroplane.

This gives a close approximation to what is called 'carefree manoeuvring', leaving the pilot to concentrate on carrying out the mission without the distraction of having to fly the machine to its limits while being careful not to overstep the mark, which inevitably uses up some of his mental capacity. The quadruplex system works on a 'vote': if one system fails, the other three, being in agreement, vote to overrule it. If a second system fails, then providing that two systems still agree, FBW still works. As a last resort, the Hornet retains direct electrical back-up to all control surfaces, and for extreme emergencies there is direct mechanical back-up to the stabilators only.

Duplicate hydraulics

The hydraulic system is duplicated, and the two systems are routed separately as far as possible. This was a direct result of experience in Vietnam with the Phantom, which had duplicate hydraulic systems running side by side. Consequently, a hit in the right (or wrong) place took out both systems. The hydraulic reservoirs contain a level sensing system which detects leaks and automatically closes the faulty section down, leaving the rest of the system fully operative.

The air conditioning and environmental control is by Garrett AiResearch, and the electrical power system is by General Electric. One safety feature is that no electrical power is needed for either fuel feed or transfer. A fire detection and extinguishing system is also incorporated: detection triggers a warning light in the cockpit. The extinguishing system consists of a pressurized container situated between the engine firewalls, and is actuated by the pilot through three entirely separate systems, one to the left-hand engine and its AMAD (airframe mounted accessory drive), with separate outlets to both engine and AMAD; a similar arrangement to the right-hand engine; and a third outlet to the APU (auxiliary power unit).

In the transition stage between the YF-17 and the F/A-18, the overall size of the aircraft was increased by some 12 per cent to accommodate 4,400lb (1,997kg) of extra fuel to meet the Navy's mission requirements, and to accommodate a larger, 28in (71cm) radar antenna, the minimum size that could meet the 35nm (64km) search range specified, in the nose. Meeting carrier requirements caused a disproportionate increase in weight, and the larger and more powerful F404 engines replaced the YJ101s. The wing loading had grown with the extra weight; the wing itself was enlarged by an extra 50sq ft (4.65m^2), and McDonnell Douglas added snags to act as additional vortex generators to the leading edges of both wings and stabilators. These were subsequently removed, as previously related.

As well as growing rather larger in overall dimensions and considerably heavier, the Hornet was designed to have a service life one-third longer than that of the two aircraft that it was to replace – 6,000 hours as opposed to the 4,500-hour lives of the Phantom and Corsair. Accordingly, a very demanding static strength and fatigue testing programme was instituted for the airframe, which by the summer of 1981 had exceeded the proposed 6,000-hour lifetime, and which was finally taken to a total of 12,000 hours, including 4,000 simulated catapult launches and the same number of simulated arrested landings.

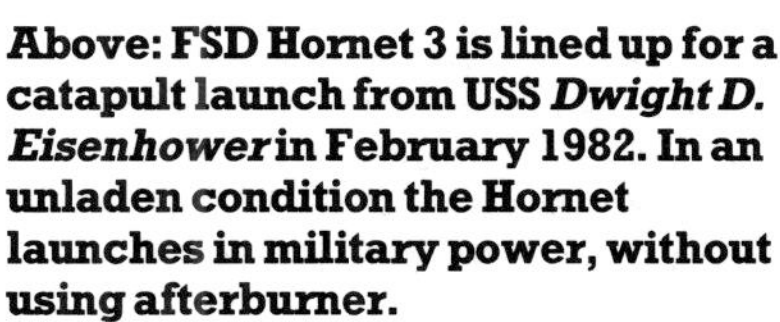
Above: FSD Hornet 3 is lined up for a catapult launch from USS *Dwight D. Eisenhower* in February 1982. In an unladen condition the Hornet launches in military power, without using afterburner.

Left: Hornet 3 apparently preparing to demonstrate a high sink-rate arrested landing, to judge by the excessively high angle of attack on the final approach to the flight deck of *Dwight D. Eisenhower*.

Static tests included maximum 11.25g symmetrical pull-ups, rolling pull-outs at up to 9g, and engine mount loadings exceeding 11g. Testing for 9g excursions was set at a rate of 27 per 1,000 hours, a far more demanding rate than the 10 per 1,000 hours stipulated for the F-15. These tests are structurally demanding, none more so than the drop test, of which 100 were carried out, at descent rates of up to 24ft/sec (7.32m/sec) on the main gear. Every fourth drop test was a simulated in-flight engagement, nosegear-first landing at descent rates of up to 27ft/sec (8.23m/sec).

It was hardly surprising that some failures occurred, but finding fault is the purpose of a test programme. If no failures were anticipated, the costly test and development programme could be eliminated. If no failures were to be found, then perfection would have been achieved, and the test programme would still have been worthwhile for the high confidence level gained. Fortunately, none of the failures was particularly serious, and the causes were eliminated fairly easily.

After approximately 300 spectrum hours of fatigue testing a fatigue crack developed in bulkhead No. 453. Investigation established that this was due to secondary bending, and the thickness of the metal in the affected areas was increased to give extra stiffness.

Much later, after 2,428 spectrum hours

Right: Main gear retraction is complicated by the positions of the Sparrow missiles. The gear retracts rearward and rotates through 90deg for stowage in their housings under the engine air ducts.

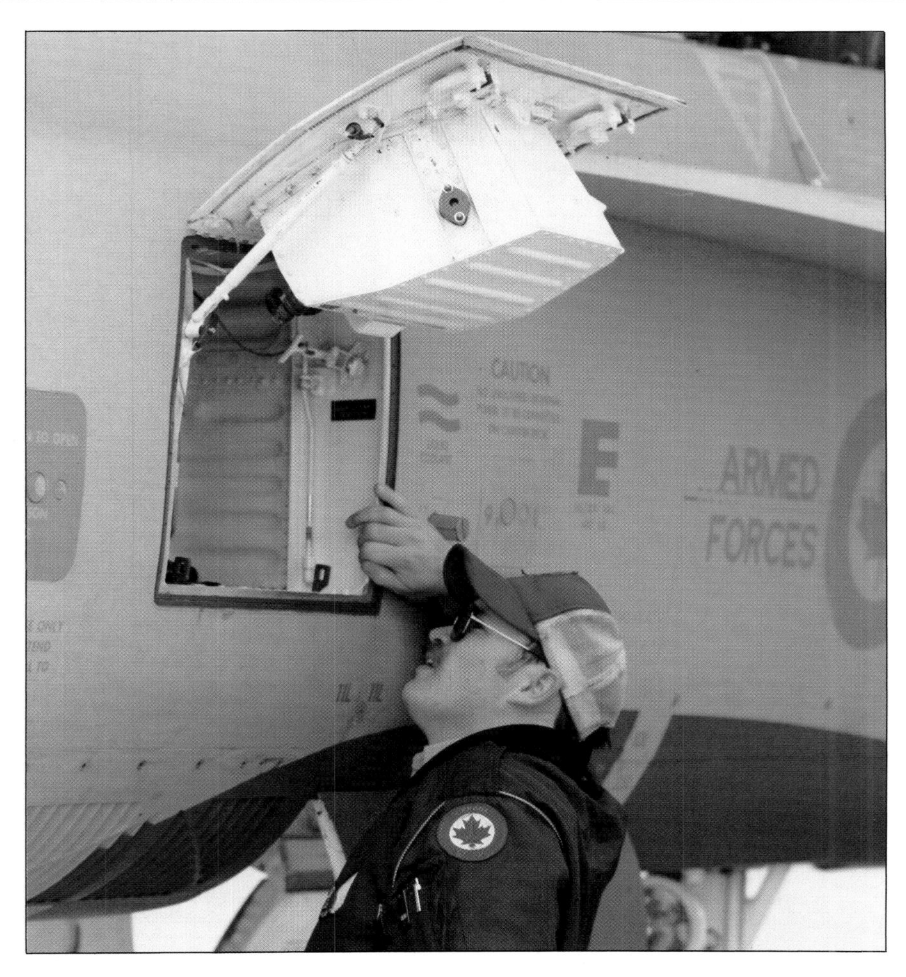

Above: The Canadian Armed Forces were the first foreign purchasers of the Hornet, with an order for 148 aircraft. A Canadian ground crewman peers intently into the gun mechanism access panel.

(approximating to 4,000 hours of aircraft life), a crack was discovered in the bulkhead which holds the main landing gear uplock system. This was caused by an error during the manufacturing process resulting from a tooling problem. A quarter-inch (6mm) diameter hole had been drilled slightly out of position. To correct this a half-inch (13mm) diameter hole had been drilled and plugged, and the new hole had then been drilled in the correct position, through the plug, creating a weakness. Only the first eight Hornets had been affected by this error before the tooling was corrected, and fatigue testing was suspended for nearly two weeks while McDonnell Douglas evaluated the use of bonding material on both sides of the affected part of the bulkhead.

At the same time, further minor cracks were found in the No. 453 bulkhead. These emanated from a group of four $\frac{3}{16}$in (5mm) diameter holes which had been drilled in the bulkhead to hold a hydraulic line swivel link fitting. In this case, the fix was to discontinue drilling these holes and attach the fitting to the bulkhead by a different method. Other minor defects included fatigue cracks found after 2,702 hours of fatigue testing in the leading edge flap drive lugs, which were redesigned to incorporate shorter pins, while a failure in the leading edge flap transmission was fixed by increasing the thickness of the wing gear, which had the effect of reducing the concentration of stress at that point. It should be pointed out that this is by no means a catalogue of disasters, but the sort of minor events that occur in any test programme.

Left: A two-seat Hornet is given a thorough wash, with intakes and AOA vanes protected by covers. Note the boarding ladder extended from its stowage in the LEX.

Below: A Canadian Armed Forces two-seat CF-18 in the hangar with the avionics bay access panels open. The avionics are LRUs and designed for rapid substitution.

One of the greatest assets of a military aeroplane is a high sortie rate: it must be able to 'get up and go' as often as possible, and not be grounded by footling defects. Reliability is a key factor in achieving this state of affairs, but equally important is maintainability, which was a fundamental consideration in the design of the Hornet. No fewer than 307 access doors are incorporated in the surface, with the selection of the type of door fasteners being based on the anticipated frequency of access. Quick release latches are used on over 53 per cent of the door areas. The approach was to obtain positive locking with no special tools needed, and no locking wire, thus reducing the ever-present danger of foreign object damage, or FOD. Almost all the access doors are accessible from deck level; only 30 require the use of work stands to reach them.

Avionics black boxes are situated at chest height and only one deep, so that extracting a failed unit will not involve the removal and subsequent replacement of a serviceable unit. The black boxes are Line Replaceable Units (LRUs) and are designed for on-the-spot substitution, so that while the old unit is pulled out and taken back to the workshop for repair a new one is slotted into its place with a minimum of fuss. The radar is track-mounted and can be rolled out for ease of access and maintenance; the windscreen hinges forwards to allow access behind the instrument panel; and the ejection seat can be removed for servicing and replaced without affecting the canopy rigging.

Both avionics and consumables are covered by Built-In Test Equipment (BITE). The engine APU can provide power for check-out of all systems without the engines running and without the need for an external power source, and a fault causes a caution light to come on in the cockpit. The pilot then calls up information on the failure on his cockpit multimode display. In the nosewheel well, a position pioneered on the F-15, where it is easily accessible to the deck crew but sheltered from the elements, is the Maintenance Monitor Panel (MMP), which pinpoints systems failures visually. Equipment is fitted with 'fail' flags which provide confirmation that repair or replacement is needed when the relevant access door is opened.

For speedy pre-flight checking, deck crews have access to a separate consumables panel which indicates critical fluid levels on a 'go/no go' basis. Without this there would have to be a time-consuming series of checks on such things as engine oil, APU oil, drive system oil, hydraulic fluid, radar coolant, liquid oxygen, fire extinguishing fluid, and liquid nitrogen for cooling the seeker heads of the Sidewinders. Servicing points are so distributed as to minimize the risk of deck crew getting in each other's way. Significant maintenance data is recorded in flight by the Maintenance Signal Data Recorder, which measures data on the engines, avionics, and structural strain gauge and makes it available at the end of the mission in the form of a print-out.

Reliability demonstration

A formal reliability demonstration was completed in November 1980, comprising 100 flight hours in 50 flights. Only 12 failures were recorded, three of which were avionics related, and only two of which affected the satisfactory performance of the mission. This compared very favourably with the Navy's requirements, which were a Mean Time Between Failures (MTBF) of 3.7 hours against a demonstrated performance of 8.33 hours, and an ultimate target of 90 per cent probability of mission success against a demonstrated 96 per cent. The mean time for repair was 1.8 hours and the direct Maintenance Man Hours per Flight Hour (MMH/FH) figure was 3.4.

Maintainability checks were held at various points in the development programme. After 1,200 flying hours the planned level of unscheduled MMH/FH was 8; the Hornet demonstrated 7.47, and at 2,500 flying hours the target of 5 MMH/FH was easily bettered, the Hornet recording 3.62. The guarantee for Fleet Supportability evaluation is 3.35 MMH/FH; this was eventually bettered by a wide margin.

The Hornet compares very favourably with other USN aircraft. In the year between July 1982 and June 1983, the MMH/FH figure for production Hornets was 2.62, while comparable figures for other machines were 4.54 for the A-7E Corsair, 5.17 for the A-6E Intruder, 5.6 for the F-4S Phantom, and 5.86 for the F-14 Tomcat. During the same period, the Hornet's Mean Flight Hours Between Failures (MFHBF) was also remarkable, production aircraft turning in a figure of 2.1 against 0.7 for the Corsair, 0.6 for the Intruder, 0.8 for the Phantom and 0.6 for the Tomcat.

Above: A contrast in styles. The enlarged LEX, canted fins and small fixed intakes are the major points of difference displayed by a two-seat F-18 alongside the same manufacturer's fourth pre-production F-15 Eagle.

A further maintenance bonus results from the fact that the Hornet is to replace both the Phantom and the Corsair. Regardless of whether the Hornet is the F-18 fighter or the A-18 attack variant, it remains essentially the same aircraft, and only 4,000 different support items need to be stocked to maintain it. An all-Hornet air wing will dramatically reduce the spares inventory, not so much in terms of total quantity of items stocked, although the demonstrated reliability and maintainability of the McDonnell Douglas fighter will do much to reduce this, but in the number of types of spares. A current mixed air wing of Phantoms (12,500 support items) and Corsairs (8,000 support items) is something of a logistical nightmare. The all-Hornet air wing will show a reduction in the number of components to be stocked of the order of 80 per cent.

The Hornet is constructed in stages, with individual sections being fabricated separately before being brought together in a predetermined order for final assembly. The rear and centre fuselage sections are constructed in Northrop's Hawthorne plant, then put together before being shipped to the McDonnell Douglas factory in Saint Louis for final assembly. The vertical fins are also manufactured in Hawthorne. The other major sections are the wings, stabilators, forward fuselage and cockpit and the landing gear.

Aluminium sheet structures are automatically drilled and routed, while heavier aluminium structural sections are formed in a gigantic two-storey machine manufactured by the Hydraulic Press Manufacturing Co, which can exert a pressure of up to 7,000 tons (711,000kg). Components which will undergo high stress, fatigue, or high temperatures, such as the stabilator pivots or engine firewall linings, are made of titanium. A process known as super-plastic forming/diffusion bonding shapes titanium sheet, which is made plastic by a combination of temperatures of 1,650deg F (900deg C) and pressures of 250psi (16.8 bars), while the world's largest profile milling shop cuts the finished components from forgings using computer controlled machines.

Pipe bending is also computerized, and coloured plastic caps on the open ends of the tubes prevent contamination prior to assembly. The electrical wiring is installed in pre-made bundles during

McDonnell Douglas F/A-18 Hornet cutaway

1 Radome
2 Planar array radar scanner
3 Flight refuelling probe, retractable
4 Gun gas purging air intakes
5 Radar module withdrawal rails
6 M61A1 Vulcan 20mm rotary cannon
7 Ammunition magazine
8 Angle of attack transmitter
9 Hinged windscreen (access to instruments)
10 Instrument panel and cathode ray tube displays
11 Head-up display
12 Engine throttle levers
13 Martin-Baker Mk 10L 'zero-zero' ejection seat
14 Canopy
15 Cockpit pressurization valve
16 Canopy actuator
17 Structural space provision for second seat (TF-18 trainer variant)
18 ASQ-137 Laser Spot Tracker
19 Wing root leading edge extension (LEX)
20 Position light
21 Tacan antenna
22 Intake ramp bleed air spill duct
23 Starboard wing stores pylons
24 Leading edge flap
25 Starboard wing integral fuel tank
26 Wing fold hinge joint
27 AIM-9P Sidewinder air-to-air missile
28 Missile launch rail
29 Starboard navigation light
30 Wing tip folded position
31 Flap vane
32 Leading edge flap drive shaft interconnection
33 Starboard drooping aileron
34 UHF/IFF antenna
35 Boundary layer bleed air spill duct
36 Leading edge flap drive motor and gearbox
37 Engine bleed air ducting
38 Aft fuselage fuel tanks
39 Hydraulic reservoirs
40 Fuel system vent pipe
41 Fuel venting air grilles
42 Strobe light
43 Tail navigation light
44 Aft radar warning antenna
45 Fuel jettison
46 Starboard rudder
47 Radar warning power amplifier
48 Rudder hydraulic actuator
49 Starboard all-moving tailplane
50 Airbrake
51 ECM antenna
52 Radar warning antenna
53 Formation lighting strip
54 Variable area afterburner nozzles
55 Afterburner duct
56 Engine fire suppression bottles
57 Arrester hook jack and damper
58 Port all-moving tailplane
59 Afterburner nozzle actuator
60 Tailplane pivot bearing
61 Arrester hook
62 Tailplane hydraulic actuator
63 General Electric F404 afterburning turbofan engine
64 Engine digital control unit
65 Formation lighting strip
66 Engine fuel system equipment
67 Port drooping aileron
68 Single slotted Fowler-type flap
69 Aileron hydraulic actuator
70 Wing fold rotary actuator and gearbox
71 Port navigation light
72 AIM-9P Sidewinder air-to-air missile
73 Leading edge flap rotary actuator
74 Port leading edge flap
75 Airframe mounted engine accessory gearbox, shaft driven
76 Leading edge slat drive shaft
77 Auxiliary power turbine
78 Flap hydraulic jack
79 Twin stores carrier
80 Outboard stores pylon
81 Aft retracting mainwheel
82 Mk 83 general purpose bombs
83 AIM-7 Sparrow air-to-air missile
84 Mainwheel shock absorber strut
85 Inboard stores pylon
86 Main undercarriage pivot bearing
87 Hydraulic retraction jack
88 Radar equipment cooling air spill valves
89 External fuel tank
90 Air conditioning system heat exchanger
91 Radar equipment liquid cooling units
92 AAS-38 forward looking infra-red (FLIR) pod
93 Boundary layer splitter plate
94 Air conditioning system water separator
95 Centreline fuel tank
96 Forward fuselage fuel tanks
97 Avionics equipment bay
98 Liquid oxygen converter
99 Nose undercarriage hydraulic retraction jack
100 UHF antenna
101 Retractable boarding ladder
102 Forward retracting nosewheels
103 Nosewheel steering unit
104 Landing/taxiing lamp
105 Carrier approach lights
106 Catapult strop link
107 Control column
108 Rudder pedals
109 Gun gas vents
110 Ammunition feed mechanism
111 Pitot head
112 UHF/IFF antenna
113 Radar equipment module
114 Formation lighting strip
115 Forward radar warning antenna
116 Radar scanner tracking mechanism

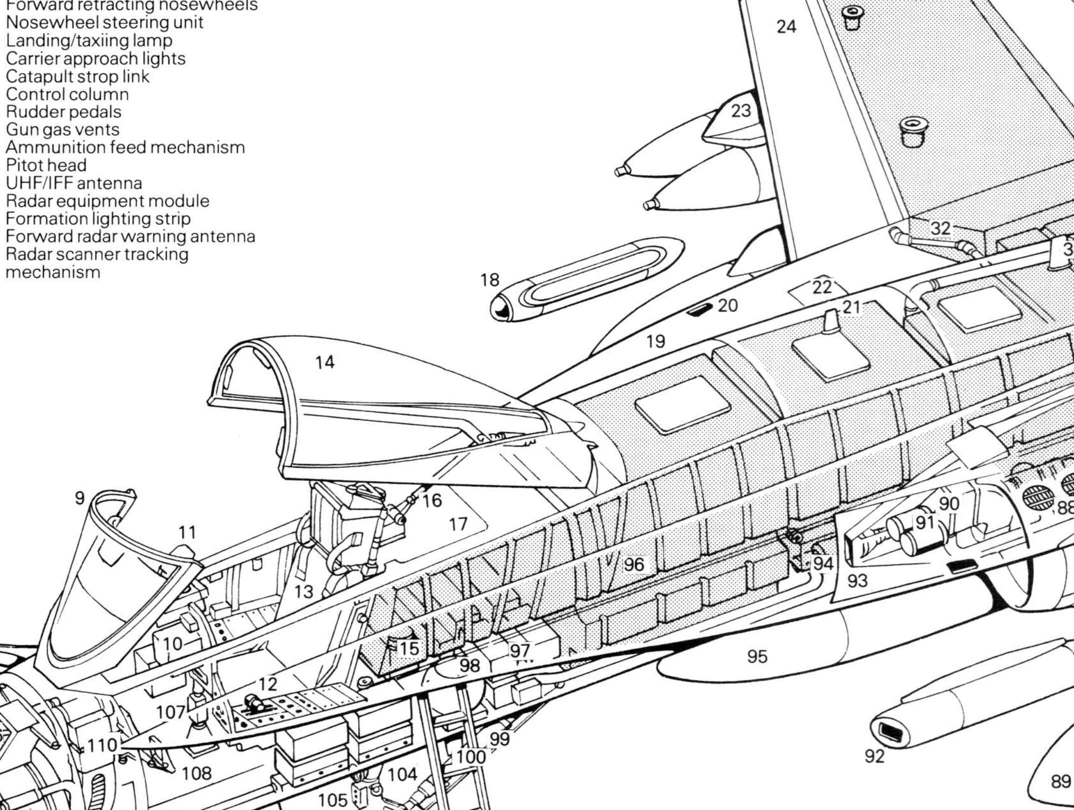

Left: Worker participation is encouraged as the 100th Hornet shipset manufactured by Northrop is officially handed over. The placards remind the workforce of their importance to the project.

Right: A TF-18A is loaded for action with two low-drag bombs on each outboard wing pylon and three cylindrical fuel tanks. AAS-38 FLIR and LST/SCAM pods occupy the Sparrow positions.

the build-up of each section, as are fuel and hydraulic lines. The fuselage, wings and empennage come together at final assembly and the systems are connected and checked. The final stages are the installation of the avionics and engines. After a comprehensive ground check, the aircraft is rolled out ready for its first flight in the hands of a company test pilot.

One production process not touched upon so far, and one which the Hornet utilized more than any other production aircraft before the advent of the Advanced Harrier, is carbon fibre composite material, in this case graphite epoxy. Composites possess high strength to weight and stiffness to weight ratios, have unique flexibility qualities and low thermal conductivity, and are extremely resistant to corrosion and fatigue. In certain applications they can be stronger than steel, stiffer than titanium, and, very significantly, lighter than aluminium. They consist of carbon/graphite or other high-performance fibres bound in epoxy resin or other matrix. In their fibre form they show near-perfect crystalline structure, and it is the parallel alignment of the crystals along the filament axis which provides the great strength and stiffness. The Hornet uses a total of 1,326lb (597kg) of graphite epoxy, giving a weight saving of 25 per cent which, coupled with strength in certain applications plus corrosion resistance, makes the extra cost involved worthwhile.

Although McDonnell Douglas has the world's largest facilities for making aircraft parts from advanced composites – more than 500,000sq ft (46,500m^2) of floor area – only 55 of the Hornet's 220 graphite epoxy panels are made there. This is partly due to the highest concentration of composites occurring on the sections made by Northrop.

Composite fabrication

The process is complex. The wing skins are the most highly stressed panels, and these have titanium inserts, with the metal bonded and tapered into the panel at the root. The sheets of composites are carefully oriented on top of one another to get the plies correct, then bonded in an autoclave with heat and pressure. To cut single ply thicknesses, McDonnell Douglas use a 1,000-watt CO_2 laser manufactured by Photon Sources Inc. which can achieve cutting speeds of between 5 and 7in (12.7 to 17.7cm) per second.

For multiple-ply cuts a reciprocating knife cutter produced by Gerber Garment Technology is used. With a 2in (50mm) carbide blade operating at 4,000 strokes a minutes, cutting rates of 7½in/sec (19cm/sec) can be achieved. Machining the panels produces fine graphite dust, which with the epoxy, can be toxic. Consequently, many operations have to be carried out under watersprays, which wash the resultant slurry into a collection point ready for disposal. Drilling composites is also an operation which needs careful judgement, and McDonnell Douglas has had to improve its drilling techniques, especially for holes through composite material/titanium/aluminium in the structure.

With all these problems, it might be thought that the difficulties of repairing battle damage in composite panels would be enormous. While admittedly it is not that easy, the following methods are believed to have been developed by the Israeli Defence Force-Air Force. Providing the damage is not so severe as to warrant a complete replacement component, the technique is to drill a hole just large enough to insert a reciprocating jigsaw, cut neatly around the damaged area to remove it, then simply insert a rubber bung! Alternatively, where this method is impracticable, composites can be patched with titanium.

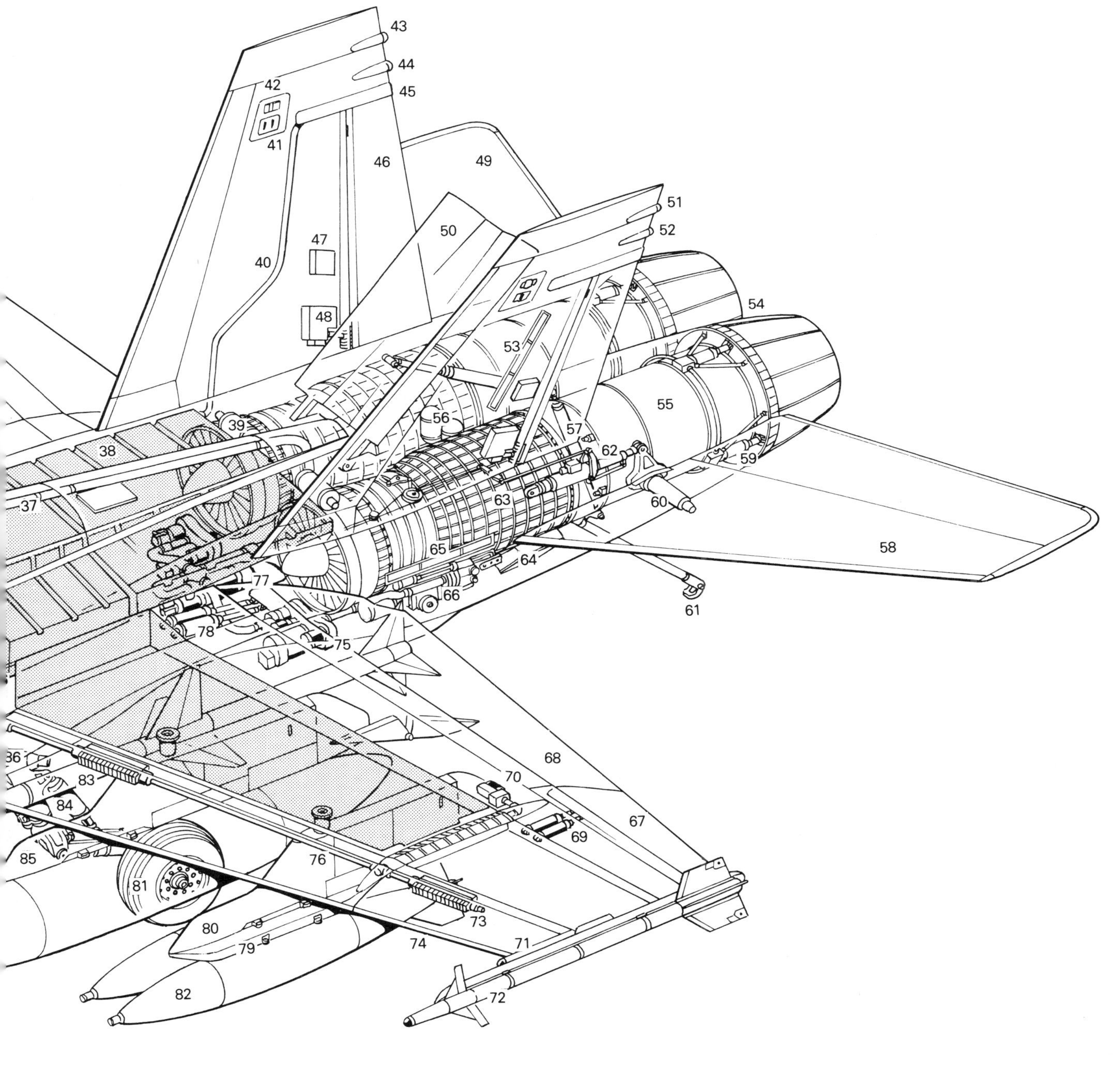

Powerplant

The evolution of the Hornet's General Electric F404 low-bypass turbofan was closely linked with that of the aircraft itself. Starting out as the YJ101, developed specifically to power the Northrop P-530 Cobra and designed for economy and reliability rather than ultimate performance, it first took to the air along with the prototype YF-17 in June 1974. When the Northrop lightweight fighter was upgraded into the McDonnell Douglas F/A-18 strike fighter for the US Navy, the engine grew with the project, gaining in size, weight and thrust to become the F404-GE-400.

Above: This head-on view of the F404-GE-400 augmented turbofan strikes the essential keynote of simplicity. From the outset, the accent has been on reliability rather than ultimate performance.

The Northrop design concept for a lightweight fighter was twin-engined from the outset. Two engines conferred a lower attrition rate and greater safety, but carried built-in penalties of their own. The structure to contain them was perforce more complex than that of a single-engined fighter, and therefore heavier, and more weight would be added by the duplication of fuel and other engine-related systems. Another penalty which is often overlooked is that two engines have twice the potential to go wrong, and double the amount of servicing and maintenance needed. Ideally then, the engine needed to be simple and easily maintainable, and to have exceptional reliability. At the same time, it had to give the high thrust to weight ratio and rapid throttle response essential for fighter operations.

The choice of engines originally lay between the Rolls-Royce RB.199 and the General Electric GE 15. General Electric agreed to develop their engine specifically for the new Northrop fighter, and the choice was made, not that it ever seemed very likely that an American aircraft manufacturer would design a new product around a British engine. Redesignated J101, the new powerplant was first seen in public at the Paris Air Show in May 1971. The GE 15 had incorporated technology from the F101 turbofan, then under development to power the Rockwell B-1 supersonic bomber, and many of the ideas were carried over into the J101.

Emphasis in design of the J101 was placed on reliability rather than ultimate performance. Cost was naturally an important consideration, and the 'design to cost' concept was treated as part of the technology of the engine. Although bearing the J prefix used to denote a turbojet, it was in fact a turbofan engine, albeit with a very low (0.2) bypass ratio. General Electric described it at this stage as a continuous-bleed turbojet, with the excess delivery from the low-pressure (LP) compressor being discharged around the core. For this reason it was semi-facetiously referred to as a 'leaky turbo-jet'.

Turbofan advantages

In pure turbojets, the afterburner and efflux nozzle is exposed to the superheated exhaust gases from the turbine. In a bypass engine, or turbofan, while some of the bypass air mixes with the core exhaust and is burned in the afterburner, the remainder is used to cool the engine external skin and nozzle, and no secondary flow for cooling the engine or exhaust nozzle is required, thereby considerably reducing complexity, drag, weight and cost. A further advantage is gained during afterburning in that the bypass air is still relatively rich in oxygen, whereas the core exhaust has already passed through the engine where much of its oxygen has been consumed.

The J101 was a physically small engine, 12ft 1in (3.68m) long and with a maximum diameter of 2ft 8½in (0.83m). With three low-pressure and seven high-pressure compressor stages, it achieved a compressor pressure ratio exceeding 20:1, and an annular combustor eliminated the smoky exhaust trails that war has shown lead to MiG pollution. It featured just two turbine stages, one high- and one low-pressure, and a variable converging-diverging nozzle. Its static thrust rating was 9,000lb (4,082kg) at full military power, and 15,000lb (6,800kg) with full afterburner, which gave it a thrust/weight ratio in the region of 8:1.

The J101 was made up of seven major modules, a feature which greatly facilitated ease of repair and maintenance. At the front was the LP compressor, a three-stage axial flow design. Variable inlet guide vanes regulated the engine air flow. Behind it came the HP compressor, of seven stages, which had been developed from the F101 turbofan. Some of the stages featured variable geometry to ensure efficient operation. Under the HP compressor was positioned the electrical-hydro-mechanical engine control module, designed to provide stall-free operation regardless of any rate of throttle movement anywhere in the flight envelope and thrust range. Between the HP compressor and the HP turbine came the combustor.

The next module in line was the single-stage HP turbine driving the HP compressor. Both the blade design and cooling in this stage were derived from the F101. Then came the LP turbine which drove the LP compressor at the front of the engine. This featured convection cooled blades, and convection cooled vane segments brazed into pairs on the nozzle. Finally there was the afterburner module, the design of which was based on that of the tried and proven J85, as used in the F-5E. This had an annular pilot flame holder, and a single-stage main fuel distributor provided smoothly modulated thrust variation.

Having the advantage of using proven technology from the F101, development of the J101 was comparatively rapid, so that initial component testing occupied just 14 months. Testing of the first core engine began in March 1972, and the first complete engine test took place during the following July. In the meantime, the USAF had issued its request for proposals for the LWF, an Air Force contract following at the end of April 1972, and the Y prefix, denoting preproduction was added. The engine thus became the YJ101-GE-100.

Simulated flight testing, carried out at the Arnold Engineering Development Center at Tullahoma, Tennessee, covered the performance envelope from high altitude to sea level supersonic speed, and various speed/AOA combi-

Above: The J101 low bypass ratio turbofan was developed to power the Northrop YF-17. It is often forgotten that for the fly-off against the YF-16 the YF-17 was using early development engines. The J101 was designed for ease of maintenance, with seven modules.

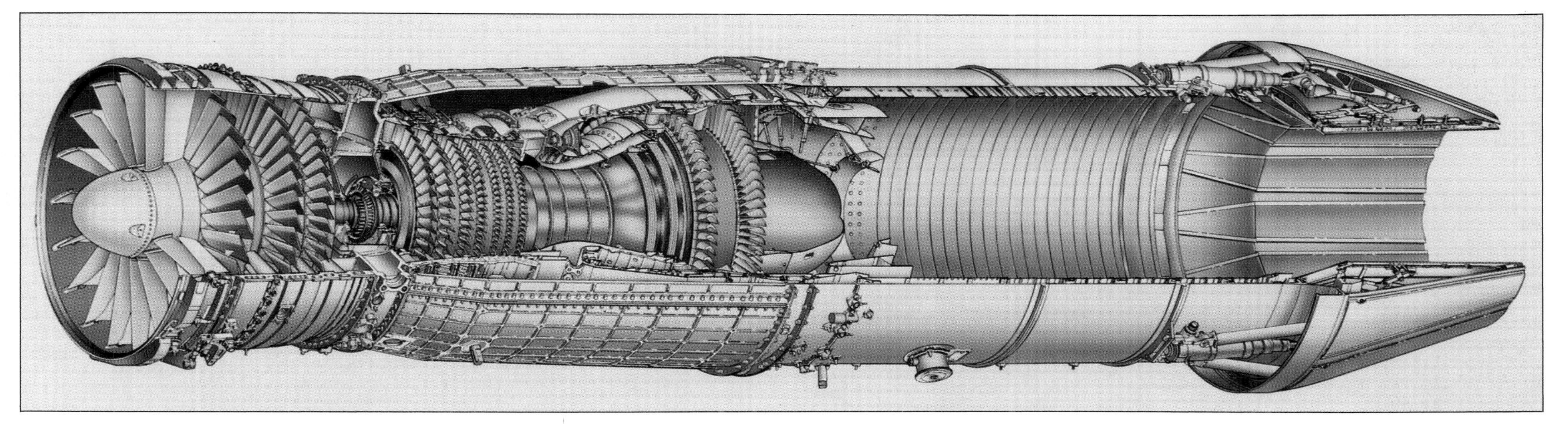

Above: Cutaway view of the F404, showing the simplicity of the layout by comparison with earlier turbojets. Accessories are mounted on the airframe rather than the engine, so it is not 'handed'.

nations. The Prototype Preliminary Flight Rating Test (PPFRT) was completed in December 1973, using a single engine, in just 101 test hours, and the USAF cleared the engine for unrestricted operation throughout the entire flight envelope. The YJ101 first flew in the YF-17 prototype on June 9, 1974.

Altogether, seven engines were used in the short YF-17 flight test programme. A total of 302 flights, amounting to 719 flight hours, were clocked up, during which the YJ101 proved to be remarkably fault-free, not one engine-related delay being recorded. Peacetime operations are considerably more arduous for an engine than those flown in war. Not only do training sorties tend to last longer, but at least one, and possibly several combats may be simulated, whereas on a war mission none at all may occur. Reliability in peace and survivability in war are the keynotes.

Below: An F404 engine on the test rig. Accelerated mission oriented testing condensed the operational mission cycles, concentrating on areas of maximum stress. Each AMT hour represents five flight hours.

It should be remembered in this context that for the ACF competition the YF-17 was using what was to all intents and purposes an experimental engine, while the rival General Dynamics YF-16 was powered by the Pratt & Whitney F100 engine already developed for the F-15. In fact, the Fort Worth company did consider using two YJ101s in their machine, but the YF-16 was a rather smaller aeroplane than the YF-17, and the weight and drag penalties of accommodating two engines were shown by design studies to be unacceptable. As we have seen earlier, the single-engined design was declared the winner of the competition.

The YF-17's two engines and larger airframe were to prove a blessing in disguise. Although the ACF competition had been lost, the YF-17 was considered by the Navy to be the most suitable aeroplane for development to meet their multi-role requirement. But to turn a lightweight fighter into a carrier-suitable, multi-mission machine was obviously going to promote it into the midleweight class, and more thrust would be needed if a dramatic and unacceptable reduction in the new fighter's performance was to be avoided.

Upgrading the J101

The obvious answer was to upgrade the YJ101, which so far had proved outstandingly successful. The result was the F404-GE-400, the F designation acknowledging that it was a turbofan rather than a turbojet, although it was at first referred to as an augmented turbojet, while the number in the 400 range denoted that the project was funded by the US Navy. The new engine was very similar to the YJ101, but scaled up by about 10 per cent and with the bypass ratio increased to 0.34, still less than half the ratio of the F100. Corrosion-resistant materials, essential to counter the salt-laden environment of carrier operations, were used throughout.

The F404, at 13ft 2in (4.01m), was 13in (34cm) longer than the YJ101, and the fan diameter was increased by one inch (2.5cm). The mass airflow was raised about 10 per cent to 140lb/sec (63.5kg/sec) and combined with a 50deg F (28deg C) increase in turbine inlet temperature, and the pressure ratio was increased to 25:1. The thrust ratio remained at 8:1. These improvements resulted in a dry thrust of 10,600lb (4,800kg) and a maximum afterburning thrust of 16,000lb (7,250kg). This put it in the same thrust class as the General Electric J79, used to power the ubiquitous Phantom among other aircraft, which could reasonably be described as the F404's predecessor.

To see how far engine technology had progressed in 20 years, a brief comparison is in order. In achieving comparable thrust, the F404 was, at 2,121lb (962kg), barely half the weight and two-thirds the length of the J79. A 25:1 pressure ratio achieved with just ten stages compared very favourably with the 13.5:1 ratio of the J79's 17 stages, while the total number of components per engine was just 14,400 against 22,000.

Above: The nozzles of the F404 are variable, with a 12-petal external cover. Here they are shown in the fully closed flight idle position.

The small size of the F404 contributed directly to the weight saving: fewer parts, only three structural frames and sumps, with their attendant lubrication systems, and just five main bearings. Yet weight reduction, although important, was not the only consideration, because the F404 could have been made lighter still. As with the YJ101, design to cost was an integral part of the programme and in several areas weight was the trade-off to keep cost down. Typically, these were the use of solid rather than hollow compressor blades, cast rather than fabricated structures, and solid metal rather than honeycomb material in casings and ducts. Steel was also used instead of titanium where the substitution was found cost-effective.

Powering a carrier fighter is the most demanding role for an engine in the entire aviation spectrum. It must be both reliable and durable, able to withstand not only the thrust and environmental changes encountered in the fighter mission, but also the repeated stresses of catapult launches and arrested landings, for which a design factor of 11g was built in. Furthermore, the deck idle thrust must be very low, so as not to cause embarrassment to either the pilot or deck-handling crew in a crowded area.

The development programme for the F404 was the most comprehensive for an engine ever. While components were originally scheduled to undergo some 5,000 test hours, in the event about 8,000 hours were clocked up, and 14 development engines underwent more than 13,000 factory test hours over a period just short of five years. The first F404 engine test took place in January 1977, a month ahead of schedule, and quickly demonstrated the required sea-level performance, and the first of six engines arrived at the Naval Air Propulsion Test Center (NAPTC) at Trenton, New Jersey, shortly afterward. Nine engines were delivered in 1978 and a further 24 in 1979. PPRFT took place in May 1978, and the first flight, in Hornet 1, in November of the same year; the Model Qualification Test (MQT) was completed in July 1979; and the first production engine was handed over in January 1980.

Engine test modes

Testing took place in three basic modes. The Simulated Mission Endurance Test (SMET) duplicated the throttle movements and power settings used in the fighter and attack missions. Three tests were held, each of 750 hours, approximating to three years of operational service. The Accelerated Mission Test (AMT) used the SMET missions as a basis, condensing the operational cycles and concentrating on the areas where damage was most likely to be caused. Each AMT hour represented five hours of operational usage, so that the 2,000 plus hours of AMT logged in the development programme represented more than 10,000 flight hours. Finally, Accelerated Service Testing (AST), the so-called Hornet Hustle, was flown by Hornets 9 and T2. The AST was a 1,000-hour programme, the first half of which was flown by McDonnell Douglas test pilots and the remainder by Navy pilots. The flight programme was very intensive: on three separate occasions, Hornet 9 flew six sorties per day.

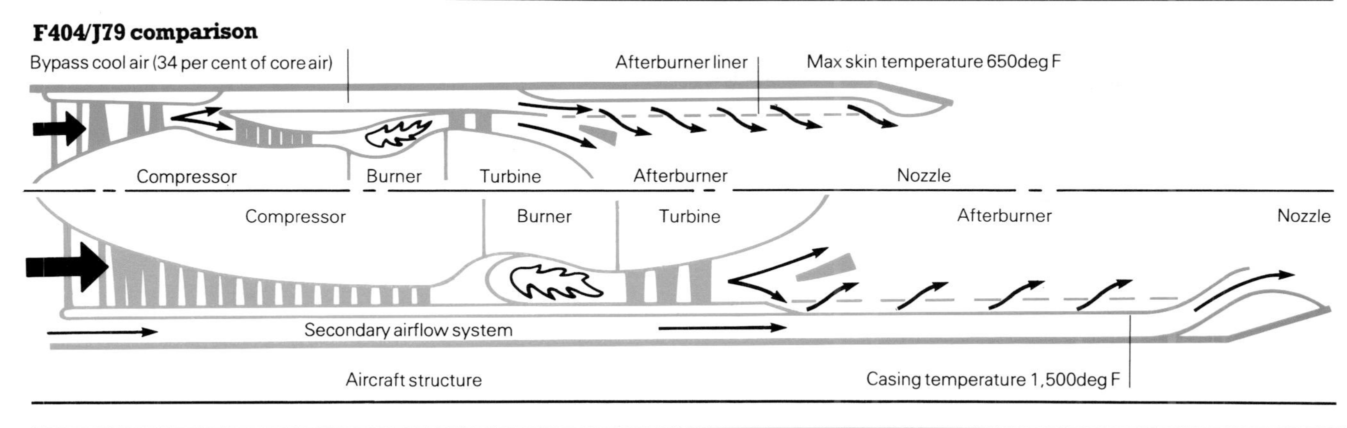

Left: Comparison of the F404 (top) with the J79, used to power the F-4 and F-104. Comparable thrust is achieved for half the weight and two-thirds the length of the earlier engine.

Below left: The auxiliary power unit (APU) gives the Hornet a self-starting capability and also provides power for systems checks from internal sources.

Below: Engine fires are extinguished by selective discharge into any of three areas: starboard engine and AMAD, port engine and AMAD, or the auxiliary power unit.

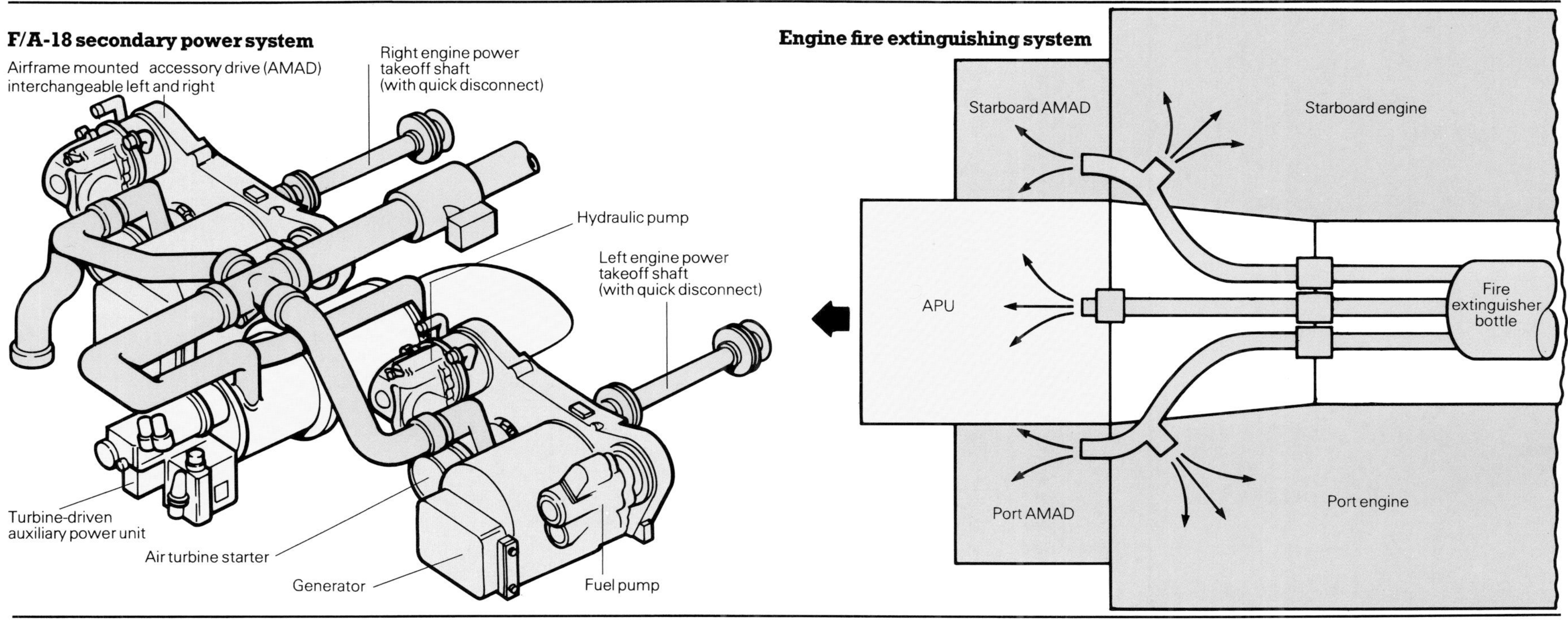

The F404 flight test programme proved remarkably trouble-free. The engine was shown to be extremely stall-resistant at high AOA and various combinations of yaw and sideslip, and no compressor stall was experienced in the first 300 flights from Patuxent River. This was in part due to the careful integration of the airframe and inlet design, particularly the LEX, with the engine. The LEX reduced the angle of the airflow into the inlets to approximately half the angle of attack. Some stalls, afterburner blowouts, and engine flameouts were later experienced at AOA of between 50deg and 90deg, but the stalls quickly corrected themselves and both engine and afterburner were found to relight automatically. The engine was also found to be wonderfully responsive, accelerating from idle to full afterburner in less than four seconds, and throttle slams from flight idle to maximum power and back were tested and found to cause no problems.

All this is not to suggest that the F404 was entirely fault-free. The specified acceleration time from Mach 0.8 to Mach 1.6 was not achieved, although the Hornet showed that it could match even F-15s in drag races up to Mach 1.2. Certain shortfalls were revealed in specific fuel consumption – the amount of fuel burned per unit of thrust per hour – but although these could only partially be compensated, this was not felt to be particularly serious. The sfc of the F404 at military power is 0.85lb per pound of thrust per hour. Other engine-related problems were a single case of a No. 4 bearing failure, and turbine blade fractures, which are described in the previous chapter.

One fault was discovered as a result of a small power loss to both engines in the same aircraft, following relight tests. With the engines stripped down, the blades of the HP turbine were found to be worn approximately 20 mil (0.5mm). Checks made on other engines revealed a wear range of between 4 and 8 mil (0.1-0.2mm). It was concluded that a temperature difference between the blades and the casing had caused rubbing and resultant wear. As the performance of both the affected engines was still above specification, no action was taken and they were re-installed.

Much more serious was the engine failure that led to the loss of Hornet T2 on September 8, 1980. The LP turbine disc suffered a catastrophic fracture and flew to pieces, causing irremediable damage. The casing of the F404 is designed to retain fractured turbine blades, which are, after all, a fairly common failure with any engine, but great lumps of a 90lb (41kg) metal disc revolving at very high speeds contained far too much kinetic energy to be stopped. Parts of the disc were not recovered, which hampered the investigation into the cause of the crash.

At that time, 33 flight test engines had been delivered. The discs in 12 of these had been formed with conventional castings, while the remainder had been manufactured by a 'fine mesh' powder metallurgy process known as hot isostatic forming developed by General Electric. In this process the raw material, called Rene 95, was reduced to a powder before being poured into a mould where it was subjected to extreme heat and pressure, the end result being known as '60 mesh'. The failed LP turbine disc was of 60 mesh, and turbine discs of this material were replaced immediately pending the results of the enquiry.

The findings were inconclusive, due to the fracture having occurred in a part of the disc that was never found, but it was thought that either a flaw in the material or a defect in the manufacturing process was responsible, and that the incident could be described as a 'worst case' event. General recommendations included redesigning all F404 rotor parts for maximum life. The turbine disc was to be strengthened, and holes formed between the existing holes to reduce stress concentration. The isostatic forming and forging process was not fully understood; in an on-going technology programme the process had been refined still further to produce '150 mesh' which tests showed to have four times the reliability of 60 mesh, which was used in all further engines.

Maintainability

Ease of maintenance is another facet of reliability. In line with the entire Hornet concept, the F404 engines were designed for maximum maintainability. One feature carried over from the YJ101 was the modular construction, which allowed entire sections to be replaced rapidly with a minimum expenditure of manhours. It will be readily appreciated that in the cramped confines of an aircraft carrier at sea, both manpower and space is at a premium.

Below: The lack of requirement for Mach 2 performance allowed simple fixed inlets to be used, with a saving in both weight and cost.

Special consideration was given to rapid engine changes. Unlike most twin-engined designs, there are no left and right engines on the Hornet; any engine can be fitted on either side. This was achieved by mounting engine accessories on the airframe instead of on the engine. The engine has only ten connections, or interfaces, with the aircraft and can be changed without special equipment being used, 'within the shadow of the aircraft'. Large engine bay doors under the rear fuselage open inward toward the centreline, exposing everything that needs servicing, all of which are mounted on the underside quadrant of the engine. The engine is changed by lowering it vertically out of the bay. With practice, a four-man team can complete an engine change in 20 minutes, although demonstration teams have often beaten this figure.

Servicing has been made particularly easy. There are no scheduled overhauls; just what is called 'on-condition' maintenance, which means putting right what needs attention. The necessity for this is established by an In-Flight Engine Condition Monitoring System, or IECMS, in which trained electrons whiz about the engine to check that all is well. Faults are displayed in the form of flags in the cockpit to warn the pilot, and read-outs for maintenance personnel. Engine removal therefore only takes place when a fault is recorded that requires remedial action, or one of the modules has reached the end of its scheduled life.

Left: The F404 was designed in seven modules for ease of repair and maintenance. The engine is suspended vertically to allow the modules to be disconnected.

Below: Engine removal only takes place when remedial action is necessary, or when a module reaches the end of its scheduled life. With practice, a four-man crew can change an engine in less than half an hour, and any engine can fit either side.

Borescope inspection

Apart from the IECMS, each engine contains 13 ports to allow internal inspection by borescope, although only nine of these are accessible with the engine installed. The F404 should require workshop maintenance less than twice for every 1,000 flight hours, and the target mean time between maintenance actions is 175 flight hours. This compares well with the corresponding figures for the J79, which are 3.1 per 1,000 hours and 90 hours respectively. The mission abort rate for the F404 is once every 2,000 hours.

Engine accessories, called line-replaceable units, are placed on the underside of the engine and are replaceable through the engine bay door. Each engine has an Airframe Mounted Auxiliary Drive (AMAD) System, which drives a fuel pump, a hydraulic pump, and a 40kVA General Electric VSCF generator.

An unusual feature of the Hornet is that the pilot can climb aboard and start the engines just as though he were in a car

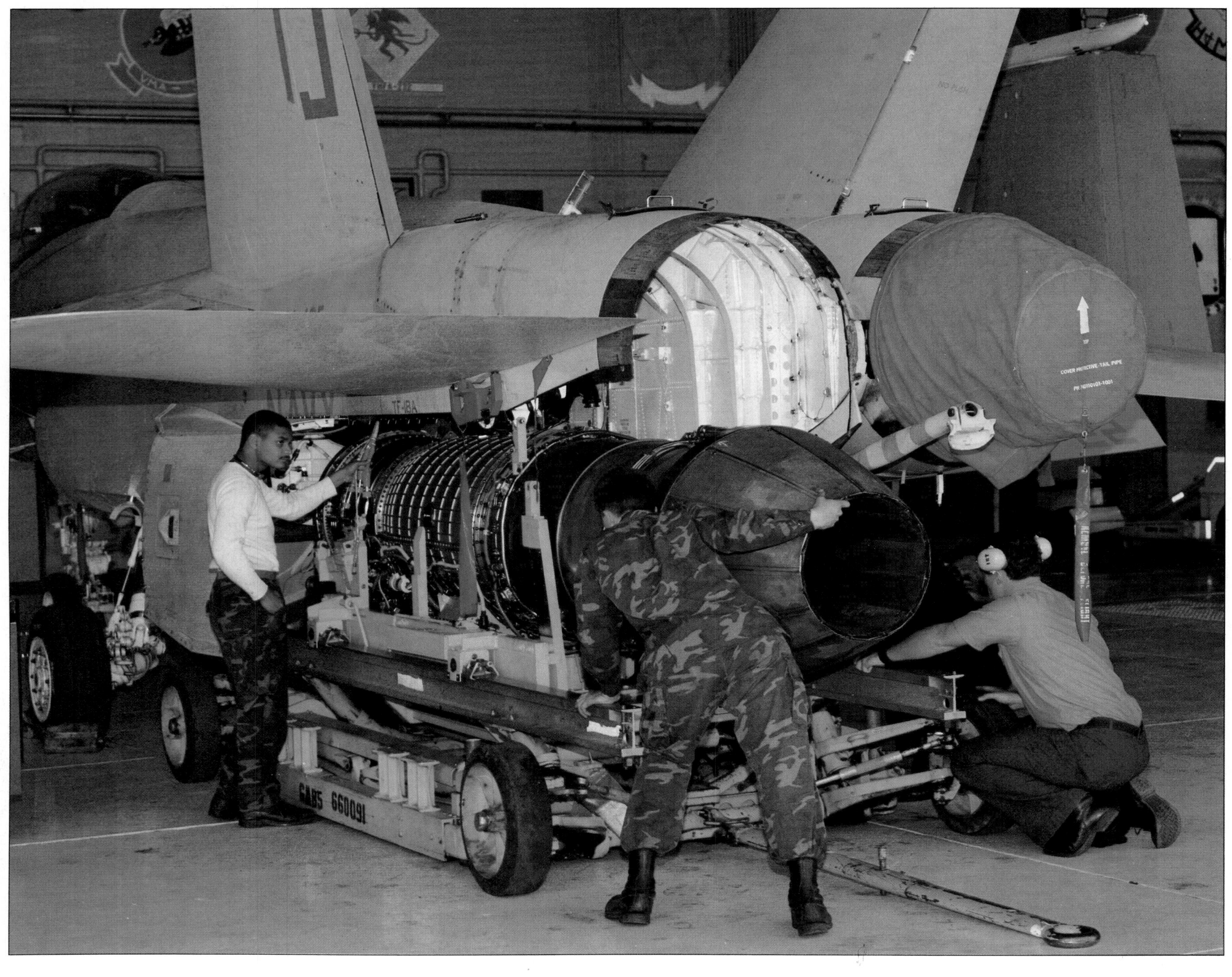

(well, almost), which saves a lot of ground support equipment, or 'yellow stuff', from cluttering up the carrier deck. The secret lies in the Garrett AiResearch Auxiliary Power Unit, or APU, which is mounted on the aircraft centreline just ahead of the twin engine bays, and is accessible through a quick-release door on the underside of the fuselage. It is a compact unit, weighing 112lb (51kg), and it develops 200hp (150kW).

The APU is started from the cockpit by battery power. It then supplies high-pressure air to the turbine starter to start the engine. Once one engine has been started, a power shaft drives the AMAD and thereby the pumps and generator, so that cross-bleed air can be used to start the second engine. Of course, the engines can always be started from an external power source if necessary.

The APU has another valuable function. By disengaging the accessory drive from the engines, all the aircraft systems can be run independently of either the engines or an external power source. This enables a full ground checkout to be made of all systems that require electrical power, hydraulic power, fuel pressure, or cooling, entirely from the Hornet's own resources. The APU can also be used to supplement air conditioning on a very hot day when engine bleed air proves insufficient for both the environmental control and avionics bay cooling systems.

Fuel system

Internal fuel is contained in four fuselage and two wing tanks, which are self-sealing and protected by foam in the wing and fuselage voids. Shaft-driven motive flow boost pumps in each AMAD unit pump fuel from the main wing and fuselage tanks to the engine feed tanks. The fuel, either JP-4 or JP-5, is supplied to the engines from separate feed tanks which interconnect for cross-feeding and are self-contained and self-sealing to provide a 'get you home' facility if the main tanks are damaged. The only fuel lines to enter the engine compartments are a main feed to each engine fuel control. This minimizes the chance of a broken or damaged fuel line spilling fuel into the hot engine compartment.

The total internal fuel capacity is 11,000lb (4,990kg) which can be augmented by three external tanks each containing 350US gall (1,324lit), bringing the maximum fuel load to approximately 17,800lb (8,075kg). A single refuelling point on the left side of the front fuselage is used for both ground refuelling and purging the fuel system, while the fuel vents and dumps are located on the top of the fins. In-flight refuelling can be used to increase the range; a retractable probe is located in a compartment at the top of the starboard side of the nose, just ahead of the cockpit.

A slight problem encountered with the Hornet's fuel system during the FSD programme was that the specification requirement of 10 seconds of normal engine operation at negative g was not being consistently achieved. The Parker jet pumps, mounted in a small reservoir designed to trap sufficient fuel for 10 seconds of negative g operation, tended to pump air pockets trapped in the fuel as readily as the fuel itself. The remedy was to replace the jet pumps with Sundstrand turbine-driven fuel boost pumps, which would not suck air.

The F404-GE-400 is the engine developed specifically for the Hornet, but other F404 variants are in the pipeline. The F404-GE-100 has been developed to give 17,000lb (7,700kg) of thrust for the Northrop F-20 Tigershark, while the F404J, with 18,000lb (8,150kg) of thrust will power the Swedish Gripen. General Electric predict that the F404 will be uprated to 19/20,000lb (8,600/9,100kg) during the next four years.

Left: Care and concentration as the engine dolly is positioned ready for the start of an engine change. Such changes are done 'within the shadow of the aircraft' as a matter of routine – an important consideration at sea.

Below: Engine accessories are placed on the bottom quadrant and are accessible through the engine bay door, as seen here. Accessories are all LRUs (line replaceable units).

Right: Flaps down for the photo call, an F-18A Hornet lights the afterburner for the benefit of the camera. The nozzles are small by present-day standards and are a direct aid to reducing the risk of detection.

Avionics

It is not enough for a modern fighter to have outstanding performance and handling qualities. It also needs clever systems for target location, weapons delivery, threat detection, navigation and communications if it is both to survive and to carry out its mission successfully. Current technology allows very sophisticated systems to be designed small enough to fit a single-seat aeroplane, and the Hornet's cockpit is a *tour de force*, presenting the information in such a manner that one man can fly the demanding spectrum of missions required of an aircraft designed for both fleet defence and long-range attack.

At the heart of the Hornet's complex avionics system is the cockpit, where all the information comes together for use by the pilot. Much of the controversy surrounding the aircraft has been based on quite justifiable doubts as to whether one man could handle all the information to be thrown at him and still fly the mission successfully. Previous Navy fighters, the F-4 Phantom and the F-14 Tomcat both had two-man crews, and they could be pretty hard-pressed at times.

The challenge to McDonnell Douglas was formidable: to take the Northrop YF-17 lightweight fighter and dress it out as a carrier-borne multi-role combat aircraft involved a tremendous increase both in the complexity of the systems and in the amount of information that would need to be presented. The Hornet had to be able to replace both the Phantom and the Corsair, and also to supplement the Tomcat in the fleet defence role; the systems and instrumentation requirement for all these tasks was enormous. Moreover, the ejection seat, raked back at an angle of 18deg as compared with the 15deg of the F-15's seat, brought the pilot's knees higher, reducing the instrument panel and console area available to only about 60 per cent of the usable area in the F-15 cockpit, but with more systems to control and display.

Starting from scratch

To design an effective cockpit, McDonnell Douglas started with the proverbial clean sheet of paper. Mission analysis was the first step. Whether flying an air-to-air or air-to-ground mission, the pilot would have up to three different air-to-air weapons, a combination chosen from more than two dozen air-to-ground weapons, all the aircraft systems and about 250 switchology functions to handle. Originally, the A-18 version was to have a moving colour map display that was not required in the F-18, while the Marine Corps wanted one UHF and one VHF radio in their aircraft rather than the two UHF sets that were to be the standard Navy fit. These differences were resolved by fitting moving map displays in all Hornets and adopting the Navy radio fit as standard.

Mission analysis identified three main workload areas: (a) weapon and sensor management during combat, in which time was critical; (b) communications, navigation, and identification (CNI) systems management throughout the entire flight spectrum, with special emphasis on carrier operations in conditions of poor visibility; and (c) systems mode management and miscellaneous requirements, which were usually not time-critical, but still occupied valuable console space as well as units of the pilot's mental capacity.

Cockpits of other aircraft were studied, in particular the company's own F-15, and work done as part of the US Navy Advanced Integrated Modular Instrumentation Systems programme was examined, as was the cockpit proposed by McDonnell Douglas for the Model 263 VFX contender. McDonnell Douglas had also been building a large simulator complex which had been extensively used during the design stage of the F-15. The simulator was heavily involved during the design and development of the Hornet, not least for the cockpit layout, in which both test and service pilots could try out proposed systems and suggest improvements.

The final solution lay in more intensive use of computer-aided controls and displays than on any previous aeroplane. The time-critical combat weapon and sensor management was achieved via the hands on throttle and stick (HOTAS) concept pioneered on the F-15. A pilot in

Above: Hornet pilots have been known to describe the cockpit as being "out of Star Wars". The displays reflected in this pilot's visor heighten the impression.

Top: The planar array antenna of the Hughes APG-63 radar is quite small, a fact made apparent by this assembly line photograph.

Right: One-man operability is the keynote of the Hornet cockpit layout. Information is called up as required on the CRT displays at the touch of a few of the surrounding buttons.

combat usually flies with his left hand on the throttle(s) and his right hand on the control column: using HOTAS, all necessary switches for weaponry or essential data displays are mounted on one or the other of these controls. The pilot is therefore able to control the necessary weapons, sensors or displays without moving his hands away from either control, and without taking his eyes off either the target or the head-up display (HUD). Management of the CNI functions was incorporated in the up-front control (UFC) panel located in the centre of the dash, while systems mode management is accomplished by switches surrounding three head-down cathode ray tube (CRT) displays.

Cockpit displays

The most striking aspect of the Hornet cockpit is the almost total lack of dials and conventional instrumentation. Prominent are the three 5in (12.7cm) square CRTs on the instrument panel. These are linked to the two mission computers and also the HUD. The HUD is of the twin-combiner type, with a comparatively wide 20deg×20deg field of view; its optics are located behind the CNI panel, and it is the main flight instrument for both weapon delivery and navigation, including both manual and automatic carrier landing modes. Data such as speed, heading, attitude, AOA, altitude, g loading, steering commands and cues for attack are projected either as symbols or in alpha-numeric format on the combiner glass and focussed at infinity, so that the pilot is enabled to assimilate the information without losing sight of the target or the carrier deck.

CRTs were chosen for the three head-down displays (HDDs) for their sheer versatility in showing different kinds of information in a small space. It is easier and quicker for the pilot to assimilate the information that he requires than from a conventional presentation because flight, sensor and weapon information are all grouped conveniently together on the CRT display. This has had the effect of allowing the conventional armament panel and more than a dozen electro-mechanical servoed instruments of dubious reliability to be deleted from the aircraft cockpit. CRTs also offer the best combination of contrast and resolution, two conflicting requirements, in bright sunlight.

Two CRTs are set high on the instrument panel; the multi-function display (MFD) on the right, and the master monitor display (MMD) on the left. Manufactured by Kaiser Aerospace, they are identical and interchangeable units: in the event of a failure of one, their functions also are interchangeable in flight and the mission would not have to be aborted. Each contains symbol generators, capable at need, depending on the complexity of the modes, of driving two or three displays, while the HUD can be driven by either. Each CRT has 20 push-button controls around its perimeter, which allow different operating modes or stored programs to be called up, and the software-programmed display processors can operate all the computerized displays. Between them, they have the added advantage of allowing the pilot to select the information he wants and present it where he wants it.

The MFD is the primary sensor display for radar attack and radar mapping information, and the digital computer gives a processed and clutter-free presentation. Also presented in an alpha-numeric format is flight information such as speed, attitude, weapon status, altitude etc. The symbology can be either cursive or TV raster (525 or 875 lines). The MMD is the primary warning, electro-optical and infra-red sensor and armament display, as well as projecting cautionary and advisory information on the aircraft systems.

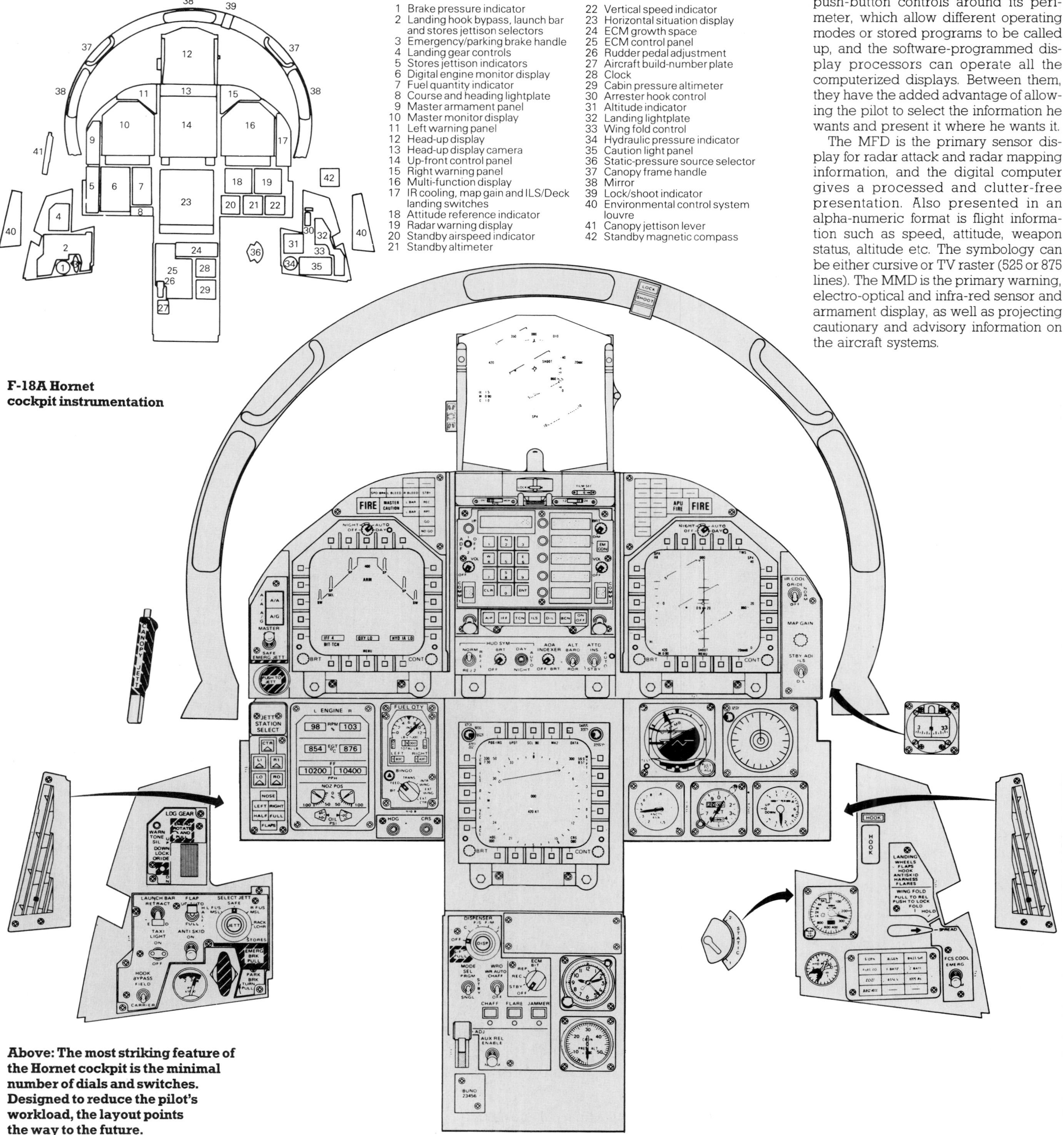

1 Brake pressure indicator
2 Landing hook bypass, launch bar and stores jettison selectors
3 Emergency/parking brake handle
4 Landing gear controls
5 Stores jettison indicators
6 Digital engine monitor display
7 Fuel quantity indicator
8 Course and heading lightplate
9 Master armament panel
10 Master monitor display
11 Left warning panel
12 Head-up display
13 Head-up display camera
14 Up-front control panel
15 Right warning panel
16 Multi-function display
17 IR cooling, map gain and ILS/Deck landing switches
18 Attitude reference indicator
19 Radar warning display
20 Standby airspeed indicator
21 Standby altimeter
22 Vertical speed indicator
23 Horizontal situation display
24 ECM growth space
25 ECM control panel
26 Rudder pedal adjustment
27 Aircraft build-number plate
28 Clock
29 Cabin pressure altimeter
30 Arrester hook control
31 Altitude indicator
32 Landing lightplate
33 Wing fold control
34 Hydraulic pressure indicator
35 Caution light panel
36 Static-pressure source selector
37 Canopy frame handle
38 Mirror
39 Lock/shoot indicator
40 Environmental control system louvre
41 Canopy jettison lever
42 Standby magnetic compass

F-18A Hornet cockpit instrumentation

Above: The most striking feature of the Hornet cockpit is the minimal number of dials and switches. Designed to reduce the pilot's workload, the layout points the way to the future.

Left (from left): The Hornet's primary cockpit displays are the master monitor display (MMD), horizontal situation display (HSD), head-up display (HUD) and multi-function display (MFD). Functions of the MMD and MFD are interchangeable, and either can drive the HUD.

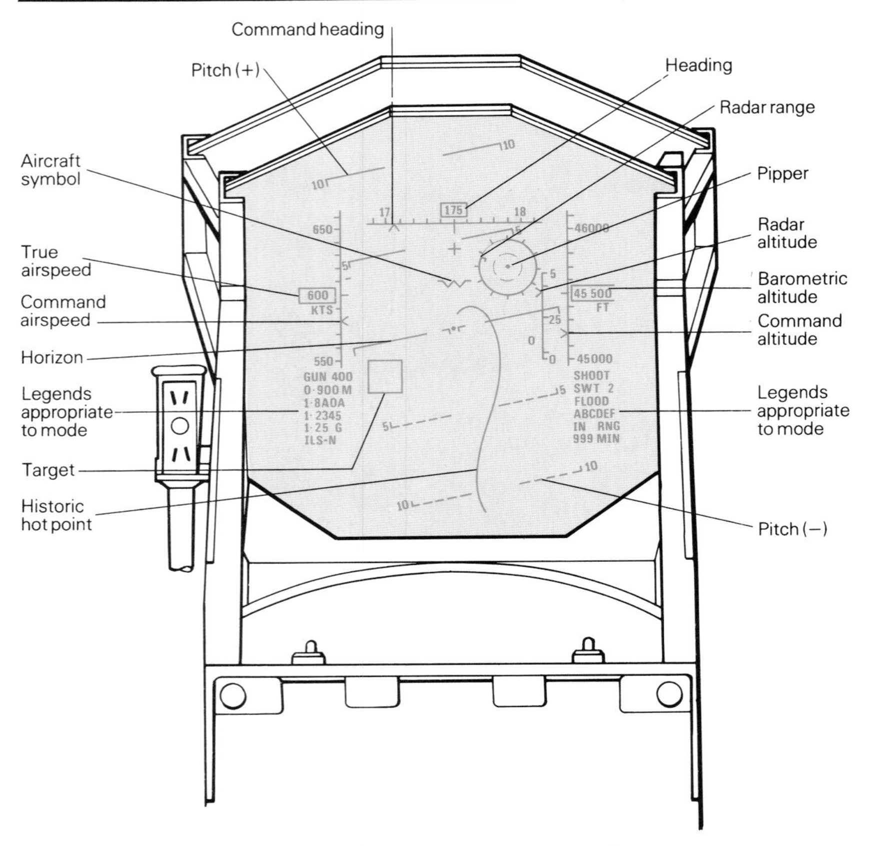

F/A-18 head-up display symbology

Left: Typical HUD symbology, showing all the information the pilot needs as he closes on his target. Among the data displayed are range, height, speed and weapon selected – in this case the gun, which is shown as having 400 rounds remaining.

At a lower level, below the UFC panel, is a third CRT. This is the horizontal situation display (HSD), based on the British Ferranti system but repackaged and licence-built in America by Bendix. It consists of a coloured, film-projected moving map acting both as a horizontal position indicator and as a display for attack information such as time/range to target, Tacan steering and INS waypoint steering commands, and it updates position as required. It also presents electronic warfare and threat indications. The HSD has push-button controls in the same manner as the MFD and MMD, which call up the information requested. An ingenious feature of the HSD is that it has a lens system that effectively forms an aperture 10in×7in (25.4cm×17.8cm) for the pilot to see the display in bright lighting conditions as though it were hooded. At night, the pilot can lean slightly forward, which effectively removes the aperture to a point outside his line of vision and prevents him being dazzled by the display. Almost every avionic system is linked to the three CRTs, as described below in connection with the specific functions.

Situated between the MMD and MFD, and below the HUD, is the up-front display, which deals with CNI functions. Supplied by McDonnell Douglas Electronics, it is so positioned that only a slight glance down from the HUD is necessary. The bottom row of buttons, reading from left to right, select autopilot, Identification/Friend or Foe (IFF), Tactical Air Navigation (TACAN), Instrument Landing System (ILS), Data Link, and Beacon, with an on/off switch on the extreme right. Just above, at extreme left and extreme right, are the switches controlling the two UHF radios. Other switches control the Automatic Direction Finding (ADF) system and essentials such as brightness, volume, etc.

The main area of the panel is taken up with a keyboard and electronic readout panels. The pilot selects a function and the readout panels display the options on that particular function; the desired option(s) are selected and the data is entered via the keyboard. The UFC panel then automatically clears, ready for further use. All controls are within easy reach of either hand, and with practice numerous CNI functions can be performed under instrument conditions.

The remaining instruments are nearly, but not quite, standard. Master warning lights are used to indicate that all is not well, but the detailed information on the malfunction appears immediately in a corner of either the MMD or the MFD. Engine and fuel state data is presented in digital form using mechanical drum counters. A departure for the Navy is the use of white cockpit lighting at night. In the unlikely event of complete power failure or loss of displays, standby instruments are located at the bottom right of the dash. They consist of pneumatic airspeed, altitude, and vertical speed indicators, and a gyroscopic Attitude Director Indicator.

HOTAS control

Not quite as way out as the displays, but still very advanced, is the HOTAS concept, which gives the pilot control of the major sensors, weapons, and displays without removing his hands from the throttle and stick, which of course is where he wants them in the heat and confusion of combat. No longer is he reduced to groping in the cockpit to find the correct switch while trying to maintain visual contact with a distant opponent.

The HOTAS system looks at first sight as though the pilot will need the manual dexterity of a concert pianist to operate the ten switches mounted on the throttles and a further seven on the stick, but experience gained on the F-15, which also uses HOTAS, plus extensive simulator tests, and, of course, the rapidly mounting flight time of the Hornet itself, has shown that the use of HOTAS lies well within the abilities of the average pilot. In this connection, it should be noted that not all the functions will be needed at once, but just a few at a time to meet the needs of the moment. Incorrect mode selection is always a potential problem, but the error becomes instantly apparent on the feedback on the visual displays, and corrections can be made almost instantaneously.

Only three of the HOTAS switches are primary to air combat; others are secondary, or are related to carrier landing functions. The primary switches are the Air-to-Air Weapons selector and the Automatic Lock-on selector on the stick, and the Target Designator control on the left-hand throttle. It should be remembered that the Hornet has two engines and therefore two throttles; while the throttles are so shaped as to be operated as a single control, there can be no guarantee that this will always be the case.

The Air-to-Air Weapons selector has three positions for selection of Sparrows, Sidewinders or guns as appropriate. The selection cues the radar automatically to the nominal parameters for the weapon chosen for range and azimuth, elevation and pulse repetition frequency. This has the added advantage of allowing the pilot to vary his search pattern by altering the weapon selected. For long-range work, Sparrow is selected and the radar automatically enters range-while-search

Below: The original Ferranti combined map and electronic display unit on which the Hornet horizontal situation display is based. A coloured film map is projected on the screen, and additional navigation information can be displayed as required.

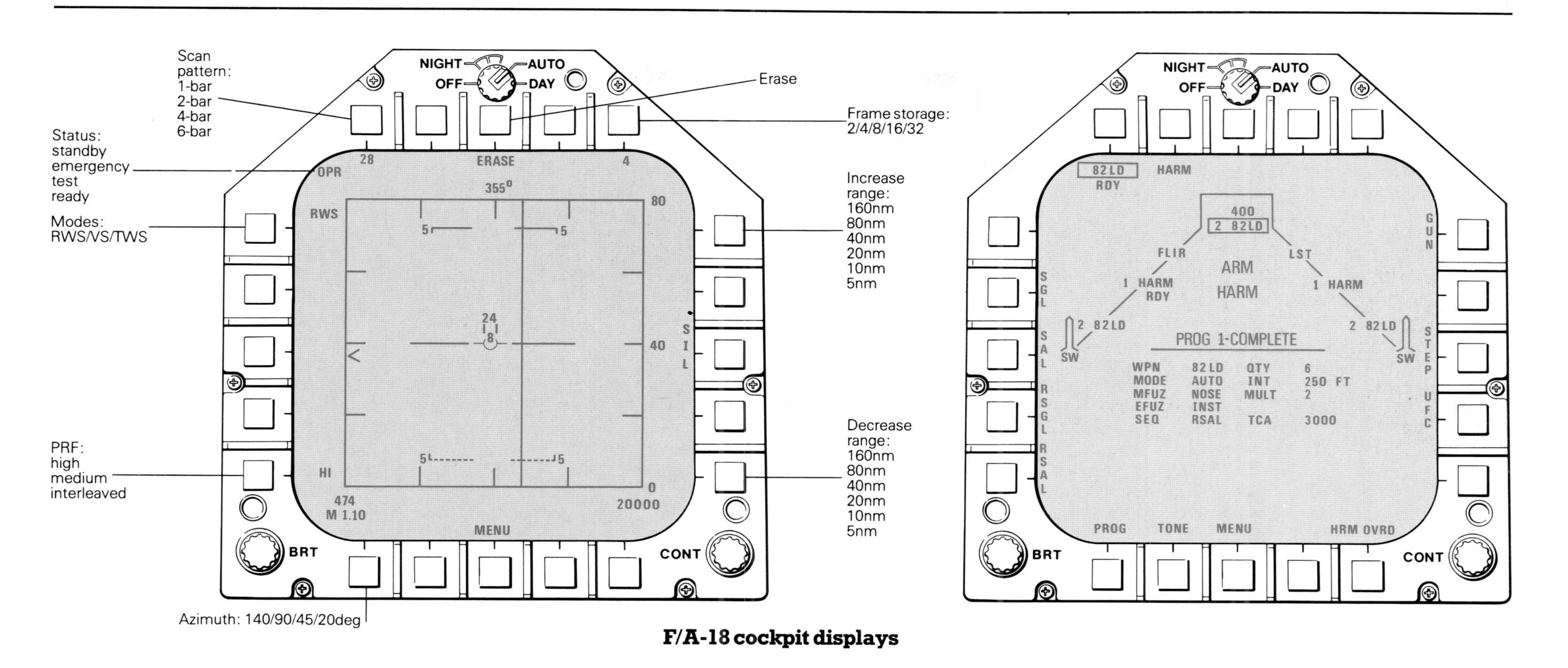

F/A-18 cockpit displays

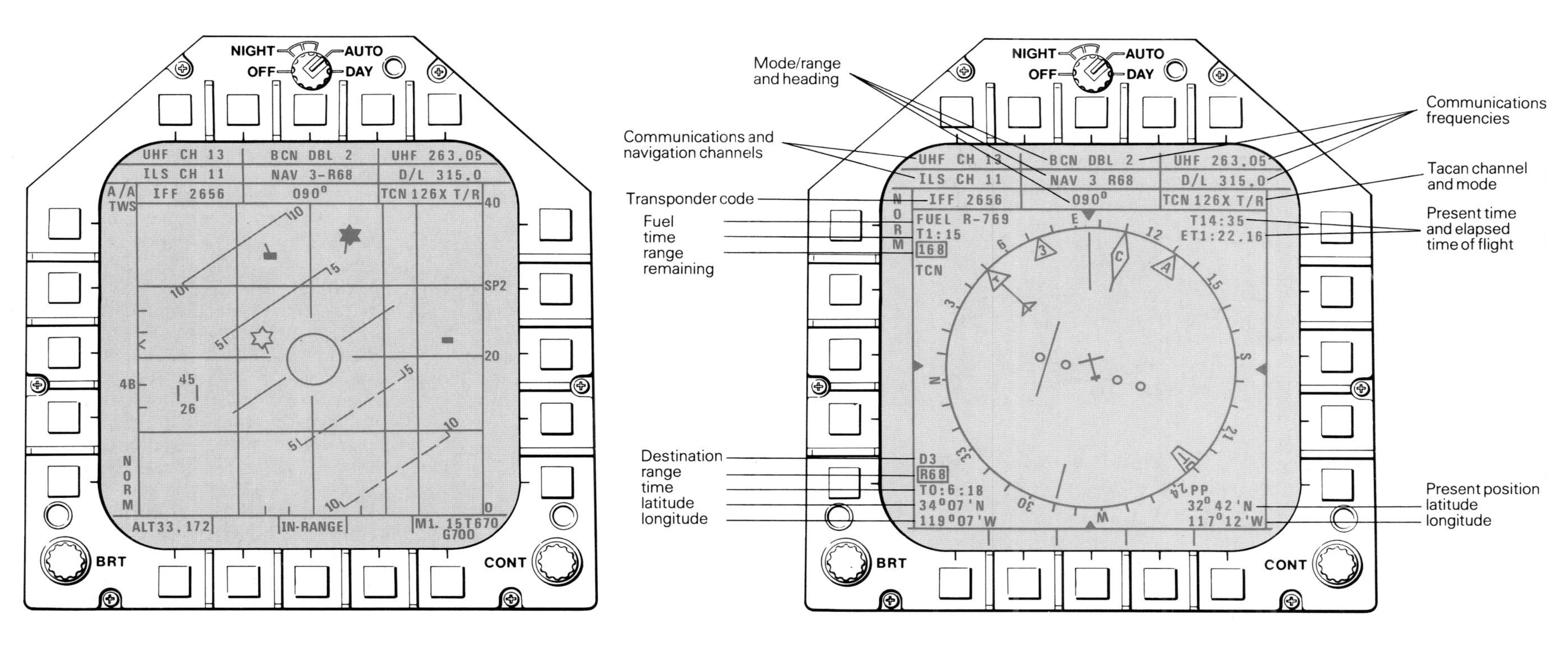

mode, out to a maximum of 80nm (147km). For Sidewinder, the search range automatically switches to 20nm (37km), with four-bar elevation scans and plus or minus 70deg in azimuth; while for guns, the search range reduces to 5nm (9km), with six-bar scans and plus or minus 45deg in azimuth. A check on the weapon selected is given on the HDDs.

The Automatic Lock-on selector is also three position and offers three modes for visual lock. These take the forms of: (1) a 3deg boresight circle on the HUD for pinpoint fly to lock-on; (2) a 20deg circle on the HUD, which gives rapid search and target acquisition within the HUD field of view; and (3) a vertical scan racetrack which opens off the top of the HUD. This is used for off-boresight lock-on, and the acquisition method used for a visual target is for the pilot to roll his aircraft until the target appears to be positioned directly above the centre of the front canopy arch. Tightening the turn then pulls the target (relatively speaking) down into the radar acquisition area, or even better, into the HUD field of view.

In all these modes target lock-on is automatic and is displayed on both the HUD and the MFD, and a 'shoot' symbol comes up on both displays when the electrons are satisfied that a satisfactory firing solution has been achieved for the weapon selected. Back-up is supplied by flashing light indicators for both lock and shoot on the top right-hand quadrant of the canopy arch. This is particularly useful when the off-boresight mode is being used, as the pilot will be visually tracking the target at a high angle-off well outside the HUD field of view in many cases.

The Target Designator Control (TDC) mounted on the left throttle is an isometric/force transducer switch which moves the designator symbol on the displays in any direction. To describe its function as simply as possible, if the pilot wishes to alter a radar mode or function, whether it be range, elevation, scan, mode, azimuth or whatever, he uses the switch to move the TDC brackets on the displays to cover whichever parameter he wishes to change, then operates the switch until the desired parameter appears. Alternatively, he can slew the brackets to cover a target symbol then, by pressing the button, designate and lock on to it. The TDC can also be used to alter the line of sight of the infra-red and laser sensors if they are carried.

Other combat-related components of the HOTAS system are the gun/missile trigger and air-to-ground weapon

Above: Typical cockpit displays. Top left is the radar display for the range-while-search mode, using high PRFs out to a distance of 80nm (147km). Top right is a stores management display, showing a bomb release programme for the six Mk 82 LD bombs; Harm and Sidewinder are also indicated. Bottom left is air-to-air track-while-scan radar mode, while bottom right is a sample horizontal situation and mission data display.

Below: Fighter pilots have always needed to fly with one hand on the throttles and the other on the control column, but as weapons and sensors grew more complex this became more difficult to achieve. The solution, pioneered on the F-15 and subsequently adopted for the F-18, is for all time-critical functions to be mounted on these controls, using the HOTAS (hands on throttle and stick) approach pioneered by the F-15.

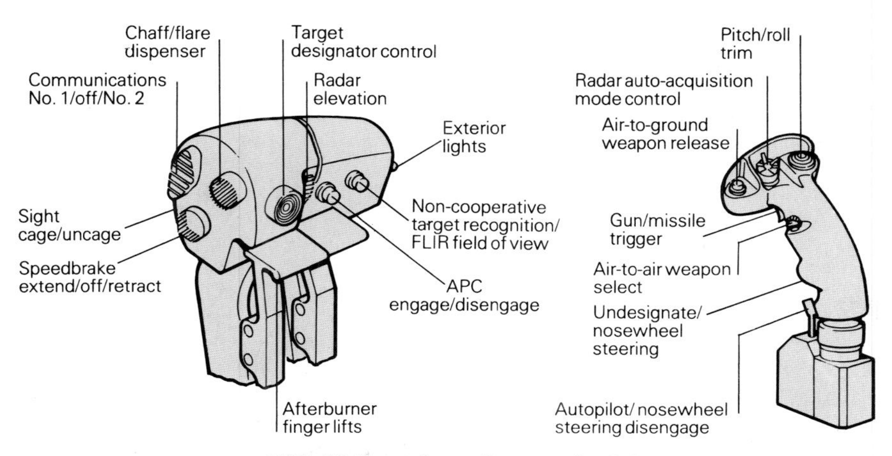

F/A-18 throttle and control stick

release switch which are mounted on the control column; the airbrake control; the infra-red seeker head cage/uncage button for the Sidewinders, which automatically slaves the IR sensor to the line of sight of the radar; a three-position communications selector switch; the three-position chaff/flare dispenser switch; and the radar elevation control, which are all on the left-hand throttle; and the non-co-operative target recognition/FLIR field of view control on the right throttle. Non-combat related functions are the autopilot/nosewheel steering disengage switch; the nosewheel steering cancel switch; and the pitch and roll trim, which are all on the control column; and the automatic power compensator engage/disengage switch (part of the automatic carrier landing system); exterior lighting switch; and finger lifts to engage ground idle power, all mounted on the throttles.

The pilot sits on a Martin Baker SJU-5/A ejection seat which is based on the tried and proven Mk 10. It provides a zero speed, zero altitude escape capability, and is effective up to 600kt (1,110km/h). Excellent all-round visibility is provided by a tear-drop shaped canopy made of laminated acrylic plastics. Visibility is possibly not quite as good as from the F-16, but there is little in it, and it is certainly good enough to make Fishbed and Flogger drivers suck their teeth. Oddly enough, two different manufacturers are involved, PPG Industries making the windshield while the canopy itself is from Swedlow.

The Hornet is more than just an aeroplane; it is a fine example of an integrated weapons system, and it is difficult to single out any one item as being particularly outstanding, especially as the degree of integration is such that almost everything seems interlocked with everything else. Having said that, the AN/APG-65 radar, manufactured by the Hughes Aircraft Company Radar Systems Group, is a fine piece of kit containing many advanced features never before incorporated in a tactical aircraft. The requirements were stringent: to produce a radar which lacked nothing in the air-to-air modes, and was equally good for navigation and air-to-ground functions; to be one-man operable and small enough to fit a medium-sized fighter; to be easily maintainable; and to have an unprecedented level of reliability, with a target MTBF of no less than 106 hours. Proposals were originally submitted by both Hughes and Westinghouse, with the Hughes design being selected at the end of 1977.

APG-65 radar

The APG-65 is a coherent pulse-Doppler radar operating in the X-band (8-12.5GHz), which is fairly standard for airborne radars as it requires a fairly small antenna – an important consideration when space is restricted, as it always is in fighters. Coherent pulse-Doppler radar dates back to the late 1950s, when the travelling wave tube (TWT) was developed.

The function of the TWT is to increase the level of power of a signal that is fed into it. Essentially, a radar sends out massive signals and gets minute ones back in return, and the more powerful the signal transmitted, the better the return. Using a signal from a continuously running coherent oscillator, the TWT produced pulses suitable for radar in which every pulse is exactly in phase with the preceding and following pulses. This enabled the Doppler shift – the observable frequency change when the range between the transmitter and the receiver is altering – to be used in radar for the first time.

One immediate advantage was that for the first time a low-flying aircraft could be detected against the ground returns, or radar echoes from the surface, since a moving target gives a shift in frequency returns which makes its echo different from the echoes bouncing back from the ground. Digital computers then sort out the echoes that are different. Most returns are likely to be shown to be moving in conformity with the flight path of the radar-carrying aircraft, and as these are in most cases the echoes from the ground, a threshold can be established, the unwanted returns filtered out, and only those which are not showing a Doppler shift which is in conformity with the flight path are presented on the radar display. These are likely to be targets.

In the air-to-air mode, the system contains two weaknesses. There is little point in detecting a horse and cart simply because it is moving, so a bottom limit must be set to the velocity of non-flight-path-conformal echoes. This is usually about 90mph (144km/h). Consequently, slow-flying machines such as helicopters can be filtered out, while very fast moving surface vehicles can be acquired. Furthermore, an aircraft flying at a right angle to the flight path is also likely to be filtered out, as its relative velocity will not exceed the threshold limit. Of course, this only applies to a radar searching downwards against the ground clutter; against a clear sky background it will not apply.

Pulse repetition frequency

A fundamental choice to be made with pulse-Doppler radar is the pulse repetition frequency (PRF). The range is wide: 100,000 transmitted pulses per second and upwards is classed as high PRF, while 1,000 pulses per second is low PRF. In between these extremes comes medium PRF, which, as we shall see, is very useful. High PRF has one great advantage: the greater the number of transmitted pulses per second, the higher the average power radiated, and the higher the average power, the greater the detection range. A high PRF waveform is also excellent at detecting a target coming in head-on with a high closing speed, but it is not so good at detecting targets with a low closure rate, such as would be encountered from the tail-on aspect with a low overtaking rate. Neither is it much good at measuring range; although a low degree of frequency modulation (FM) can be impressed on the pulse as a sort of identity tag, ranging information gained in this manner is not very accurate.

The inaccuracy inherent in measuring range in the high PRF mode stems from the short time lapse between each pulse. It is difficult to tell which pulse has engendered which echo, and ambi-

Left: The APG-65 radar runs out on rails for ease of access. The antenna uses electric drive, and the WRA modules are apparent.

Below: Much of the flight testing with the APG-65 radar was carried out by this specially modified T-39D Sabreliner. The picture at left is also of this aircraft.

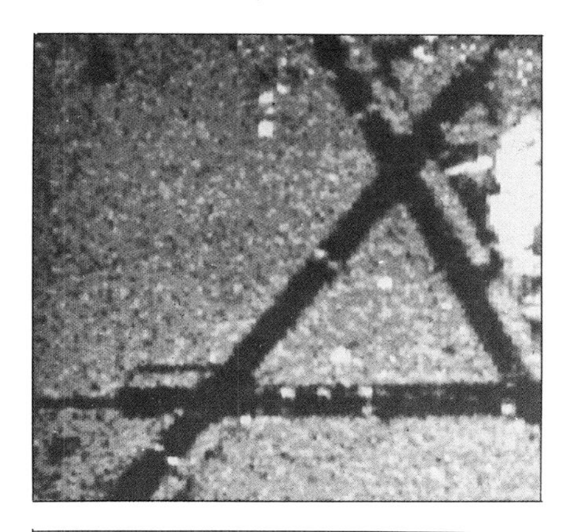

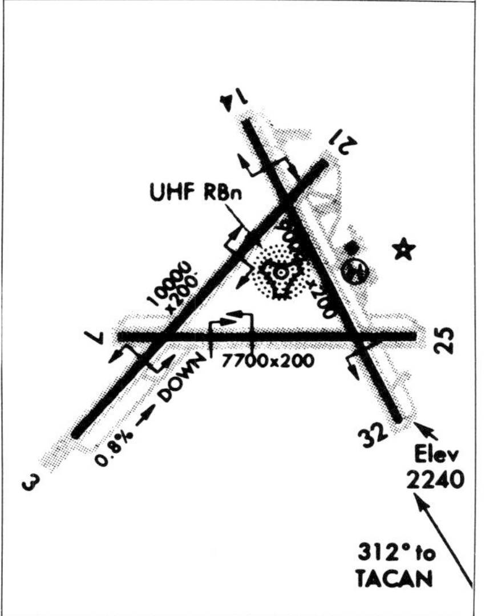

Above: Doppler beam sharpening techniques give excellent ground mapping resolution. Compare the upper picture of a DBS patch mode map with the airfield layout below it. Computer techniques are used to give a vertical picture.

guities arise in consequence. Low frequency PRF has much better ranging capability, the time lapse between pulses enabling a return to be received from a considerable distance before the next pulse is transmitted, which removes the ambiguity of high PRF.

A compromise solution of medium PRF was first used operationally in the F-15 radar, the APG-63. Medium PRF confers many advantages in the medium-range detection and accurate tracking of small, high-speed targets, the accuracy being sufficient to enable data for weapons delivery to be processed. In the APG-65, a medium PRF waveform is interleaved with high PRF. The medium PRF used is not a constant waveform, but a series of PRFs in the medium band. This in practice gives good average solutions to the problems posed by the varying velocities of different targets. The PRF variation is accomplished by the programmable gridded TWT.

The use of rapidly varying PRFs was made possible by the use, for the first time on a production fighter, of a programmable signal processor, which has the staggering ability to perform up to 7.2 million operations per second; what are called real time calculations. This allows incoming echoes to be sampled and analyzed to adjust and set the processing boundaries. Range gate and filter configurations are pre-programmed on software, unlike those of the APG-63, which give a fixed choice of selections. On the

Right: The USM-469 Radar Test System for the APG-65 radar installed at NAS Lemoore, California.

Below: The WRA modules in the APG-65 are designed for speed and ease of replacement in the field.

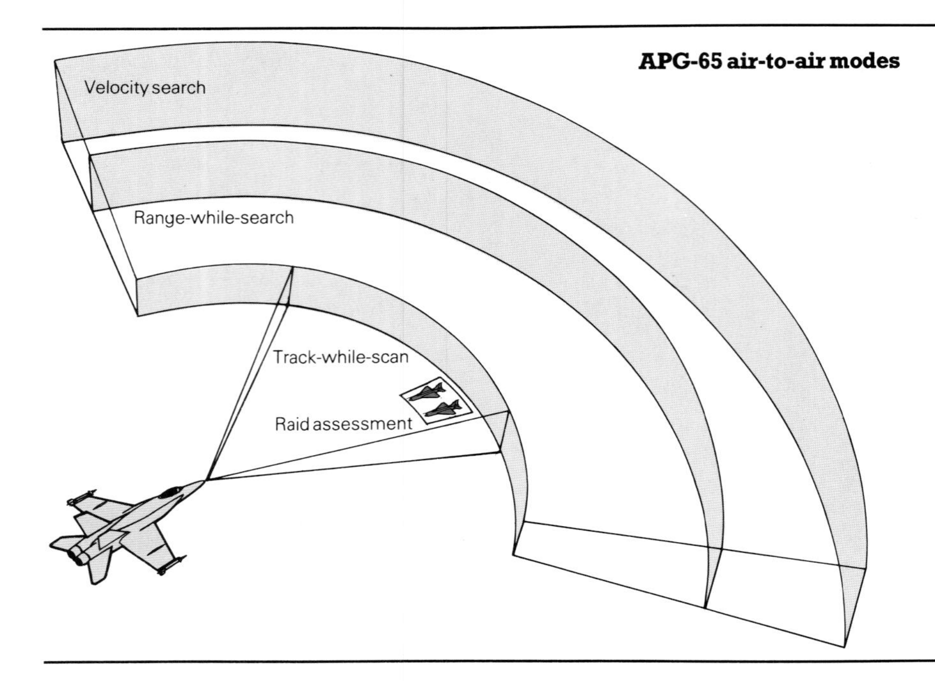

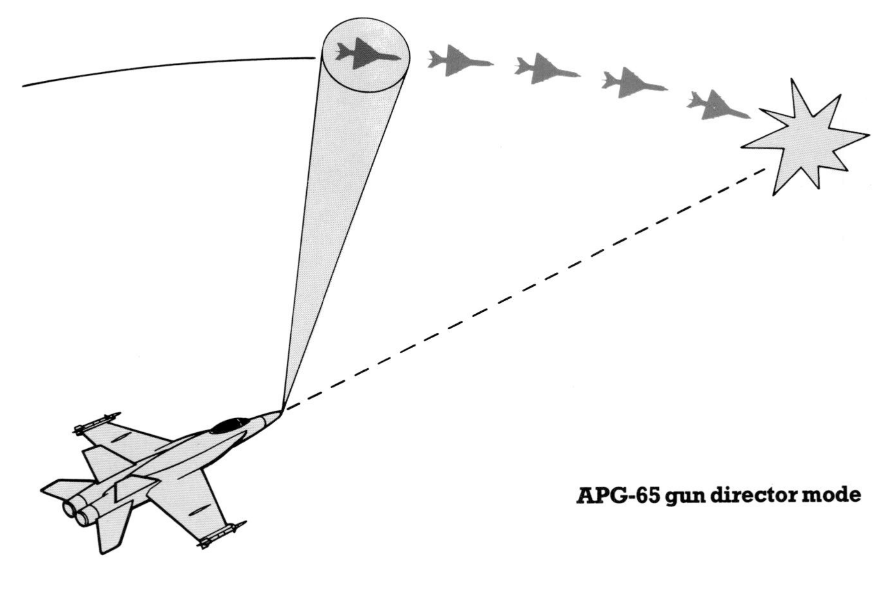

Hornet, if new modes are needed due to the requirements of new weapons or changes in the nature of a threat, or existing modes need modification, a software change will suffice.

The APG-65 is, considering its capability, remarkably small. It weighs just 340lb (154kg) excluding the rack, and its volume, excluding the antenna, is only 4.49cu ft (0.127m^3). It contains approximately 14,000 parts, compared with the 27,000 parts of the AWG-9 on the Tomcat, and on test it has exceeded its MTBF guarantee of 106 hours, which compares remarkably well with the in-service figure of 8.2 hours for the AWG-9. Modular for easy maintenance, like so much of the Hornet, it consists of five primary subsystems which are all known as weapon replaceable assemblies (WRAs).

Radar servicing

As in all the other systems, BITE is incorporated, and can detect 98 per cent of potential or actual failures and indicate in which WRA the malfunction exists. Any WRA can be substituted in 12 minutes. One great advantage of the WRAs is that they are all digital, and need no special alignment or adjustment when being fitted. For servicing, the dielectric radome (by Brunswick) swings open to the right, and the APG-65 can be run out on an extending track; the radar is fully operable even with the track extended.

The antenna is a fully balanced, low side-lobe slotted waveguide planar array, using direct electric drive, thus saving the weight and complexity of a hydraulic system. The transmitter contains the X-band gridded TWT, which alone among the WRAs is liquid-cooled to reduce both the necessary voltage and the thermal stresses, both of which reductions contribute to reliability. All other WRAs are air-cooled. The radar data processor stores instructions for the different operating modes on a floppy disc unit with a 256K 16-bit word capacity; on demand, the instructions are transmitted to a 16K capacity solid-state memory, which controls the operation of the radar.

The digital signal processor is the key element in the radar: without its real-time handling of the masses of incoming information, the entire sequence would fail. The receiver/exciter unit converts the incoming signals from analog to digital form; it consists of low-noise field effect transistor (FET) amplifiers, which give great reliability for low cost, a low-noise exciter with multiple channels, and the analog/digital converter.

The APG-65 carries out a great deal of work on its own programs, but the net results still have to be presented to the pilot on the cockpit displays, partly with symbology and partly in alpha-numeric form. Gone are the days when the raw data was presented on the CRT in analog form, and the pilot or a second crew member had to exercise a great deal of expertise in deciphering what it all meant.

Above: Velocity search detects long-range closing targets; range-while-search detects all-aspect targets; and track-while-scan follows ten targets, displaying eight.

Above right: Gun director mode uses pulse-to-pulse frequency agility to track targets and set the correct lead for a gun attack.

Right: Three air combat manoeuvre modes are available, all of which provide automatic lock-on to the first target acquired. Top is boresight acquisition, centre is vertical acquisition, used against either a higher or a turning target, while bottom is head-up display acquisition, which covers the area directly ahead of the HUD and locks on to the first target detected. A 'step-through' facility is provided to allow the pilot to reject targets successively until he acquires the one he wants.

We have seen that the APG-65 is a very sophisticated piece of kit: precisely what can it do? Its mission modes fall into three basic categories, air-to-air, air-to-ground, and navigational functions, which also contain some capabilities that we have not so far examined. Air-to-air modes are:

Velocity search. This mode utilizes high PRF for long-range detection. As we have seen, high PRF works best at long range on rapidly closing targets. The priority for this mode is early detection of targets that are likely to pose a threat within a timespan measured in minutes, rather than those that are heading in an entirely different direction and will not become a threat unless a radical change of course is made. The information is presented to the pilot as azimuth and velocity only, in other words the direction the target is coming from, and how fast it is approaching.

Range while search mode uses both high and medium PRF waveforms to detect targets at all aspects and relative velocities out to about 80nm (150km) range. The high PRF pulses are FM coded for ranging while the medium PRF utilizes the range gate filtering incorporated in the PSP. The purpose of range while search is to detect anything out there, regardless of aspect, heading, velocity, or threat potential.

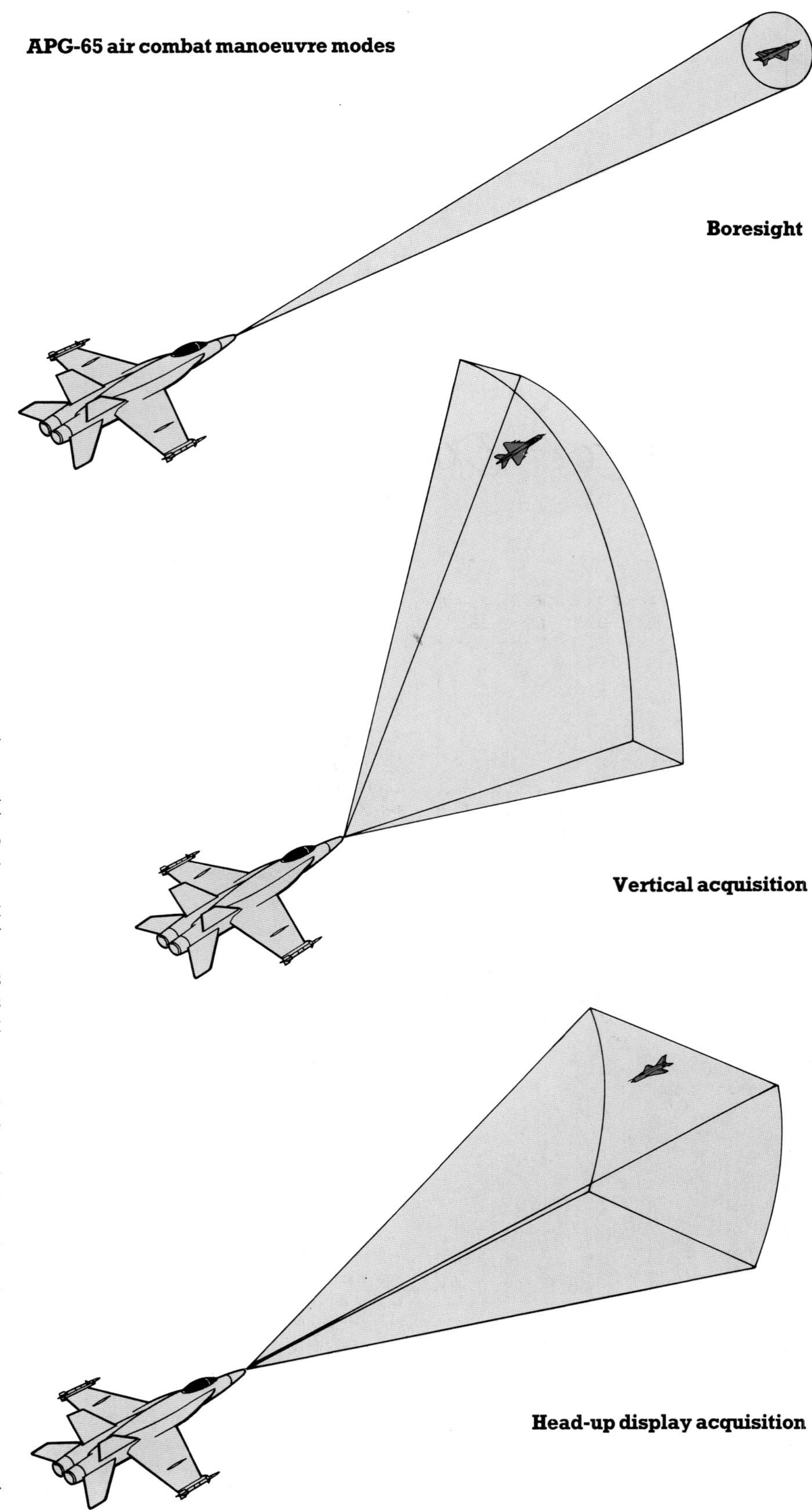

Right: The terrain avoidance mode coupled with precision velocity update confers a first-pass blind strike capability in the attack mode.

Far right: Doppler beam sharpening improves ground mapping; sector mode uses a 19:1 sharpening ratio, while patch mode uses a 67:1 ratio.

Below: High-resolution radar mapping modes greatly simplify navigation as well as target location and identification.

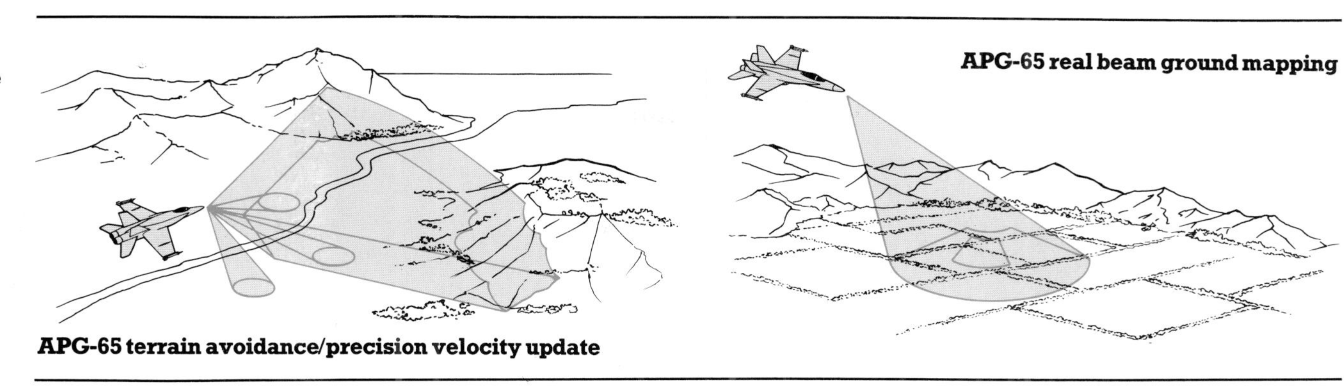

APG-65 terrain avoidance/precision velocity update

APG-65 real beam ground mapping

F/A-18 radar navigation and attack

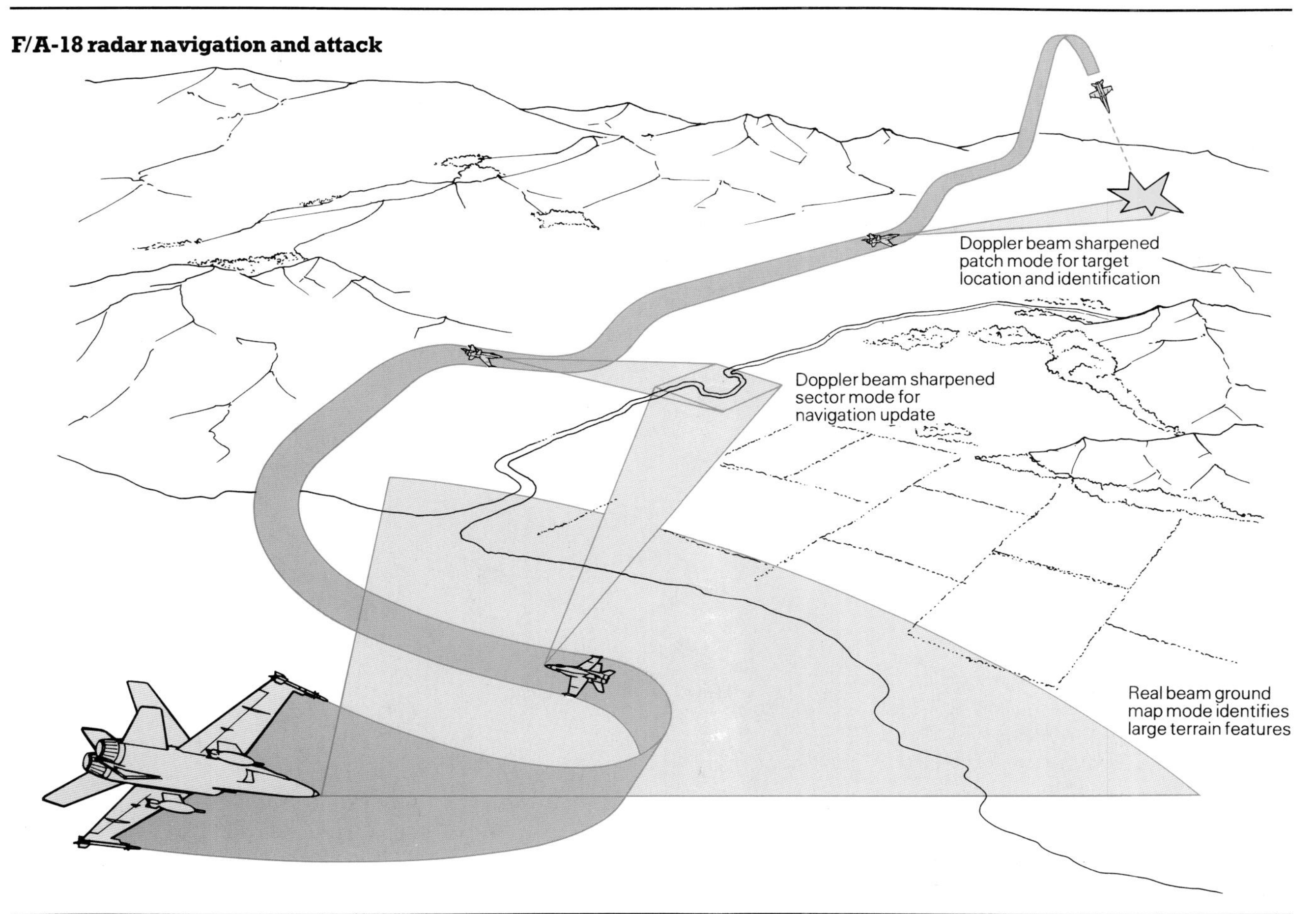

Track while scan is a medium PRF mode used during the closing phase at ranges below 40nm (75km). While not in the same league as the Tomcat's big AWG-9, the APG-65 has the more than useful ability of maintaining files on the tracks of ten separate targets, while displaying the eight most likely ones to the pilot, eight being considered the maximum number, in terms of workload, that the pilot can cope with. The aspect, altitude, and speed of the highest priority target, i.e., the greatest threat, is also displayed.

At some future date, the Hornet is to be equipped with the launch and leave AIM-120 AMRAAM (Advanced Medium Range Air-to-Air Missile). When this happens, it will provide the Hornet with the capability of engaging separate targets simultaneously.

Single target track mode is automatically cued on the HUD when the target comes within range while the radar is in the range while search mode. The pilot then switches to STT mode, which uses two-channel monopulse angle tracking able to follow the target through most manoeuvres, although computer logic is needed to extrapolate 180deg turns and Split-S and Immelmann type manoeuvres by the target. Provided only that the antenna gimbal limits are not exceeded, the radar should not break lock in the STT mode. Attack steering commands and data for weapon launch are shown on the HUD, while the velocity, aspect and altitude of the target are displayed on the MMD.

In addition to target tracking, the system continually computes launch parameters, and 'shoot' cues are displayed when a firing solution has been achieved. A special high pulse rate provides illumination for the semi-active radar homing (SARH) AIM-7 Sparrow missiles. As the range reduces to below 20nm (37km) the pilot has the option of using the heat-seeking Sidewinder, and uncaging the seeker head with the throttle-mounted switch slaves the IR sensor to the line of sight of the radar. A visual check of which target has been acquired, essential in a confused tactical situation, is provided on the HUD.

Raid assessment mode has been developed to solve the perennial problem of hostile aircraft flying in such close formation that radar discrimination is insufficiently sensitive to be able to separate them. In consequence, they appear on the display as one target. This is a matter of particular concern in the NATO defence area, and also the Middle East, where such 'bunching' tactics were brought to a fine art in 1973. In essence, the raid cannot be hidden, but the single blip gives no clues as to the composition of the force, and therefore allows no intelligent guesses to be made as to its intentions until the moment it splits up, which is generally far too late for effective counteraction to be taken.

The raid assessment mode, effective at ranges of up to 30nm (55km), provided that the enemy formation has a minimum separation between aircraft of about 500ft (150m), uses Doppler beam sharpening techniques based on expanding the area around a single target return to give increased resolution, which in turn should allow the radar to separate the individual components of a formation.

Air combat manoeuvre modes

These break down into three forms.

Boresight utilizes a narrow 3.3deg beam placed on a target which is within the boresight axis, or centreline, of the radar-carrying aircraft, the time-honoured method of pointing one's nose at the enemy, although in a manoeuvring engagement this is really of most use in the traditional pursuit attack from astern.

Vertical acquisition scans an arc 5.3deg wide by 60deg above boresight and 14deg below, once every two seconds, and is most useful for tracking a target when either the tracking aircraft or both it and the target are in a hard turn. The pilot rolls the Hornet into the same plane of motion as the target, positioning the target above the centre of the canopy arch. Vertical acquisition is most useful when both aircraft are turning hard with the target less than 60deg angle-off.

Head-up display acquisition is the third air combat mode. This scans the 20deg by 20deg field of view of the HUD, which is plus or minus 10deg in azimuth, and 14deg above boresight to 6deg below in elevation, once every two seconds.

In all these modes, which can be used over ranges varying between 500ft (150m) and 5nm (9km), the radar locks on to the first target acquired automatically, with visual cues indicating lock and shoot appearing on the CRT displays, the HUD, and via flashing lights on the canopy bow. Despite the automatic acquisition of targets, the pilot can always reject them in turn until he reaches the one he really wants; alternatively he can designate the target with the moveable cursor.

Gun director mode can be used for ranges of less than 5nm (9km) and the radar provides position, range and velocity data on the target, which drive the gun aiming point, or pipper, on the HUD. Glint, or erratic changes in the apparent radar centre of the target, which can under some circumstances move off the target altogether, is overcome to a large degree by the use of pulse-to-pulse frequency agility. This method also provides very accurate data for lead-angle prediction, simplifying high angle-off shooting; the pilot places the pipper on the target and presses the trigger. A conventional sight is used as back-up in the event of a malfunction.

The air-to-ground modes are no less impressive. In particular, the long-range surface mapping, using high resolution modes never previously incorporated in a tactical aircraft, is outstanding. To identify large geographical features from long distance, necessary, for example, when approaching a hostile coastline, the **real beam ground mapping** mode is used. This combines low PRF with pulse compression to confer long range, and non-coherent pulse to pulse frequency agility to avoid glint. The mode provides a rather crude small-scale radar map of the terrain ahead, from which large features such as river

Right: Pilot's eye view of the vertical acquisition mode. A target is seen (left) off to the right and turning. The pilot rolls his aircraft (centre) into the plane of motion of the target by positioning his aircraft in such a way as to make the target appear to be above the centre of his canopy bow. Acquisition should be automatic, but if a firing solution can not be achieved he tightens the turn, causing the target to appear to move down into the HUD. In this mode, the 5.3×74deg arc is scanned every two seconds.

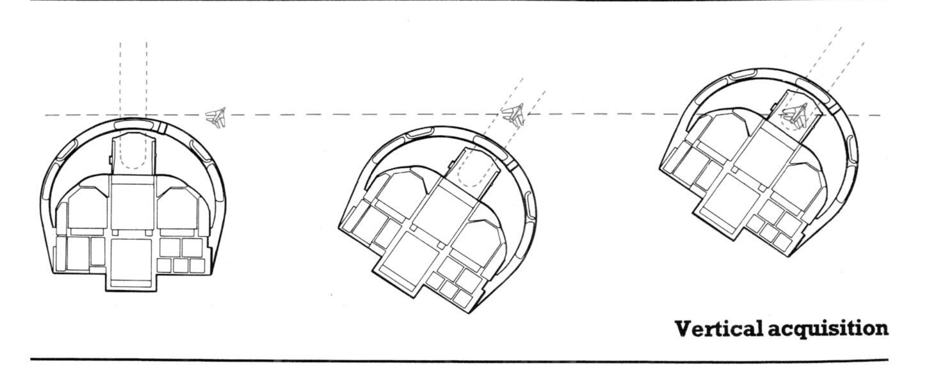

Vertical acquisition

estuaries can be readily identified. In all the ground-mapping modes, the display presentation is computer-adjusted to present the map from a vertical viewpoint rather than from the shallow angle obtained from the aircraft, which would give a distorted view and make recognition of features much more difficult. Other ground mapping modes give better resolution over smaller areas by using doppler beam sharpening (DBS). In the **DBS sector** mode, a beam sharpening ratio of 19:1 is used, while the **DBS patch** mode utilizes a 67:1 ratio.

Terrain avoidance mode is used for low-level strikes in poor visibility. An automatic terrain-following system would of course be far better but this is not a built-in capability; instead, terrain avoidance shows the pilot where the ground is, and it is then up to him to avoid it. Two sets of data are presented; one is the ground profile along the velocity vector of the aircraft (the direction in which it is travelling, which is not necessarily the same as the direction in which it is pointing), while the other shows the ground profile at a preset level below the direction of travel. Obstacles projecting through this preset level of clearance are clearly shown on the displays, which allows avoiding action to be taken. In a dive the terrain along the direction of travel is displayed, but in a climb the display shows the terrain parallel to the ground. This prevents the pilot from levelling out too soon in the event of there being a peak ahead of him.

Precision velocity update is another radar capability. It can be used to provide the Doppler input to the computer for weapon delivery, and also to improve navigation by updating the inertial platform of the INS for velocity errors. It also provides for in-flight alignment, although the demonstrated accuracy of the Litton ASN-130 INS, at 0.5nm per hour (far better than the specification requirement), is such that little correction is likely to be needed.

A carrier fighter perforce spends much of its time flying over water, and anti-shipping strikes are part of its function. To detect ships, **sea surface search** mode can be selected, although tracking is not a feature of this function. Radar clutter from the surface of the sea varies considerably according to the sea state (how rough it is). When the sea surface search mode is selected, the radar first samples the sea state. This is then analyzed by computer and a filter threshold is established to filter out the background clutter and present only those returns that do not conform, which are likely to be ships.

Other air-to-ground modes are concerned with ranging and attack. Either fixed or moving targets may be attacked, using two-channel monopulse angle tracking combined with coherent frequency agility, and ranging on designated targets is accomplished by one of two methods, depending on whether the depression angles are large or small. For large depression angles, split-gate range tracking is used, and monopulse tracking is used if the depression angles are small. A designated target is automatically acquired in this mode, which can also be used to provide ranging information when the target is designated by laser or infra-red means.

The outstanding reliability of the radar is no accident: from the outset it was a primary requirement, since it is obviously of little use to have the most capable radar in the world if it spends half its time in the repair shop. Simplicity of design was primary, as was an intensive test programme. Comprehensive testing during production is also used to detect potential or actual faults (infant mortality is the manufacturer's term) with both high and low temperature and exacting vibration conditions an integral part of the highly automated tests.

The specified requirement of 106

Right: View from the seat of the Sperry Operational Flight Trainer. Simulation is playing an increasingly important part in training pilots on new types. The realism here is excellent, apart from the fact that the pilot is not wearing gloves.

hours MTBF was met in June 1983, a whole year ahead of schedule. At this time, the requirement for the development stage achieved was only 85 hours MTBF, but as Hughes Program Reliability Manager Terry Rostker explained: "After we met that requirement, we decided to continue the test at our own risk to show that the radar can meet the mature system requirement of 106 hours MTBF (equivalent to 148 hours of failure-free running in the test chamber). At 149 hours, we stopped the tests to analyze the results and to review them with McDonnell Douglas and the Navy. Then an additional 54 hours of test were run to validate the highly successful demonstration."

The tests were stringent, to say the least. Each radar was sealed in an environmental chamber and subjected to a series of nine-hour operational cycles. Each nine-hour segment consisted of 90 minutes of cold soaking at −65deg F-(−54 deg C), followed by a further 90 minutes at −40deg F (−40deg C). The set was then switched on and allowed to warm up for just six minutes, before being continuously operated for six hours at temperatures of up to 160deg F (71deg C).

In line with the overall maintenance-free concept of the Hornet, the APG-65 requires no regular maintenance inspections, calibration, or adjustment. Only when a malfunction manifests itself does it receive any attention.

Radar Test System

An essential back-up to the APG-65 is the USM-469 Radar Test System (RTS), also designed by Hughes. When a fault is located through the Hornet's BITE, the affected WRA is removed from the aircraft and a fresh one installed. The faulty unit is connected to the RTS, whose operator keys in a test programme which is then run automatically. The problem is identified and presented on a visual display unit (VDU) which is part of the RTS, and if necessary a print-out is given. The faulty component is replaced by the operator, and the test programme is re-run as a check. The RTS also has a complete self-diagnostic facility which enables it to check itself to ensure that it is functioning correctly.

The APG-65 has automatic electronic counter-countermeasures (ECCM) built into all the operating modes, and the set is designed to resist hostile jamming. Frequency agility plays a large part in this capability but some of it is managed through the software. An advertised advantage is that the software can be rapidly reprogrammed to meet changing threats. From a practical point of view, this can be done quite easily, but it does contain a problem from the com-

Above: Air combat is simulated in the Hughes Weapons Tactics Trainer, in which the fledgling Hornet pilot spends about 50 hours. Consisting of two domes, each containing a Hornet cockpit, the WTT enables trainees to fly against each other or the computer.

Left: The first night landing is one of the less dull moments in a Navy pilot's life. Simulated carrier landings and takeoffs, coupled with emergency procedures, ensure that the pilot is as well prepared as possible.

Right: Like all USN and USMC combat aircraft, the Hornet is equipped with the ALE-39 countermeasures dispenser system. A total of 30 rapid-bloom chaff, flare and jamming cartridges are carried in each of the two dispensers mounted just behind the inlets on the underside of the Hornet's fuselage.

mand point of view. The question must arise: who decides the priorities? Is it the higher echelons of command who, with masses of data at their disposal for analysis, make a considered, global judgement, then alter all modes in a standard way for all units; or should it be the commander on the spot, who has just become aware of a major threat change and wishes to counter it before tomorrow morning's strike is launched? The question is complicated by the fact that not only would the aircraft program need alteration, but the RTSs would also need adjustments in order to be able to continue their checking functions.

Central computers

The total computer memory requirement of the Hornet was 741K 16-bit words at the outset, with room for growth. This is almost half as large again as the capacity of the F-15, which has some 500K words at its disposal. In total, the Hornet contains more than two dozen computers, which fall into two basic regimes, sensor-related and mission-related. The sensor-related computers are concerned with detection and navigation, while the mission-related computers concentrate on weapon delivery and display management.

At the heart of the Hornet's avionics are two Control Data Corporation AYK-14 mission computers, each with a 64K core memory capacity, one of which handles navigation and support while the other concentrates on weapon delivery. Should one fail, and battle damage is always a possibility to be reckoned with, the other can provide sufficient backup to give a 'fight your way home' service. They also control the multiplex bus system which consists of three channels, each containing two multiplex buses, one of which is active while the other is redundant. Two of the channels connect with other avionic systems, while the third connects the two mission computers.

Clever avionics are essential to the operational capability of the Hornet, and the software, with its own development programme running parallel to that of the aircraft, is a subject on its own. It is, for example, used to auto-release air-to-ground weapons in situations where the pilot of a Corsair would be busy hitting rows of switches on a console. The same software principle applies to the navigation and communication systems, the Automatic Carrier Landing System (ACLS), the Litton Industries ASN-130 Inertial Navigation System (INS), etc.

Threat warning is provided by the Itek ALR-67 Radar Warning Receiver (RWR), which is able not only to locate but also to identify the threat by checking its emission characteristics against a memory bank, then presenting it on a cockpit display. Electronic countermeasures are also built in. Full ECM details are naturally not available, but it is to be expected that the standard forms such as barrage jamming, range-gate stealing, and other deception jamming measures are incorporated. Once the receiver detects the presence of a hostile radar which appears to present a threat (which is half the battle), its emissions are then compared with those of known threats and identified. Further processing reveals which emission appears to pose the greatest threat, and countermeasures are automatically instituted. Passive countermeasures are also fitted, in the form of the Tracor ALE-40 flare and chaff dispenser.

The pilot must be kept warm (or cool) in his cockpit, which must also be

Above: In the attack role the Hornet carries the AAS-38 FLIR pod on the port Sparrow station. Using thermal imagery, this produces a picture on the cockpit MMD of terrain or targets at night. Definition is remarkably sharp, as evidenced by this series of pictures of a US Navy amphibious assault ship.

Above: The FLIR pod in position on FSD Hornet 7. The optical head rotates and swivels automatically to follow a designated target.

pressurized: failure to keep within the set limits would lead very quickly to pilot failure. Much the same principle applies to the avionics, and many failures are caused by high operating temperatures. Avionics cooling lies within the field of the environmental control system, and the original specification for cooling air to the avionics required it to be delivered at a temperature of 62deg F (17.7deg C) at altitudes up to 30,000ft (9,150m), with a proportionate decrease at 42,500ft (12,950m) to zero degrees F (−17.8deg C).

In practice, these limits were improved considerably, and the temperatures of avionics microcircuit junctions ran at up to 95deg F (35deg C) cooler than their specified operating temperatures, making a direct contribution to systems reliability. Another factor which directly improved reliability and helped to keep operating temperatures down was the fact that many components were derated, some operating on as little as 15 per cent of their design power level. For example, the radar has a range performance comparable to that of the much larger APG-63, yet it operates at less than half of the peak transmitter power level.

FLIR and laser pods

In the attack role, two avionics pods can be carried in the positions otherwise occupied by the two Sparrows. These are the Ford Aerospace AAS-38 forward-looking infra-red (FLIR) pod, and the Martin-Marietta Laser Spot Tracker/Strike Camera (LST/SCAM). The FLIR pod flight testing on the Hornet was completed by December 1981. It provides a round-the-clock bad weather attack capability by providing the pilot with a picture of the terrain over which he is flying in conditions too bad for visual target acquisition.

To locate the target the pilot initially selects a field of view 12deg×12deg, along a line of sight which can be varied between 30deg up and 150deg down. The pilot also has the freedom to rol up to 540deg either way before the system gives up and sulks. The area thus surveyed is presented at apparent actual size on the MMD by thermal imagery, the equipment receiving infra-red impulses (it should be noted that infra-red is just another electromagnetic wavelength, slightly above the level of visual light) and converting the energy into a televisual format electrical impulse. In this form it is displayed on the MMD as a picture of the object being observed, with remarkably clear definition. A function is also included to ensure that the image displayed in the cockpit appears the right way up, i.e., as the pilot would normally see it.

When a target is identified, the field of view can be closed down to 3deg× 3deg, which gives an image magnification of about four times, although ×10 magnification has been experimented with. The auto-tracker can then be engaged, and the AAS-38, which is integrated with both the avionics and the mission computer, presents accurate data for the calculation of weapon release solutions. The optical head can also see the target after the aircraft has passed over it, giving it the capacity for assessing the weapon's effectiveness, and in the event of the target being a bridge or other fixed, known location the FLIR pod can be used to update the navigation system.

The AAS-38 is small, 13in (33cm) in diameter, 6ft (1.83m) long and just 340lb (154kg) in weight. Like much else on the Hornet it is modular (with 10 components), easily maintainable and has a very high MTBF. It is planned to expand the AAS-38 to incorporate a laser designator/ranger in the pod, and provision has been made to accommodate the transceiver, pulse-forming network, and necessary power supply.

The LST/SCAM pod has a dual function. It searches for, acquires and then tracks laser pulse-coded energy from a pre-designated target, thus giving a bad weather/visibility first pass strike capability. It allows the target to be acquired from beyond visual range, and provides for speedy acquisition during a manoeuvring, defence-evading approach. Once the target is acquired, the LST passes data to the mission computer, which in turn presents a visual indication of the target's position, on the HUD. At the same time, aiming and weapon release information is presented.

The SCAM part of the pod uses a 35mm Perkin-Elmer panoramic camera. Directed by the mission computer, this covers the target area before, during and after the attack. The LST/SCAM pod is longer but narrower and lighter than the FLIR pod, being 8in (20cm) in diameter, 7ft 6in (2.29m) long and only 162lb (73kg) in weight.

Below and below right: The LST/SCAM pod occupies the starboard fuselage station. The crewman is fitting the pod's WRA-203 centre section.

Armament

Like its insect namesake, the Hornet carries a vicious sting. For the air-to-air role it has Sparrow and Sidewinder missiles and a 20mm Vulcan cannon, and it will eventually be equipped with the launch-and-leave AMRAAM. For the air-to-ground mission it can carry a wide variety of ordnance to match the task in hand, from laser-guided weapons to old-fashioned iron bombs, while for anti-shipping strikes specialized weapons such as Harpoon are available. While none of these is unusual, the Hornet really scores as a true multi-role fighter, rather than a fighter dressed out for alternative tasks.

Above: Close-up of the large and drag-inducing twin store rack and pylon as armourers practice bombing up a Hornet at sea. The bombs do not yet have fuzes attached.

Despite the return to fashion of the aircraft gun, missiles are the primary weapons of the modern fighter in air-to-air combat. Missiles have a chequered history, and have been the source of many misconceptions and misunderstandings. Before examining the weapons relevant to the Hornet, perhaps we should take a brief overview of the entire subject.

For a start, the term 'guided missiles' has found wide currency in the argot of aviation. This is a misnomer. Very few missiles are in fact guided, which term implies positive control by the firer, and the air-to-air missiles carried by the Hornet certainly do not come into this category. Homing missiles, or target-following missiles would be a more accurate description.

To digress briefly, missiles that can manoeuvre to follow their targets seem to have caused a certain amount of confusion in the United States in their early days, as the first manoeuvring air-to-air missile in the world to enter service, the AIM-4 Falcon, was originally allotted an experimental fighter designation as the XF-98. There was, of course, some justification for this; Hughes Aircraft had for all practical purposes created a small pilotless kamikaze aeroplane!

A myth widely disseminated in the early days of homing missiles was that they would obviate the need for manoeuvring combat between fighters. Future air encounters were to be fought at long range, with the victory going to the side which detected the enemy earliest, got into position first, deployed the longest-ranged weapons and nad the best countermeasures. War in the air was to become a war of technology: the new missiles would manoeuvre unerringly in tracking down their targets, and they could not be out-run except at the very limit of their range, and then only in the unlikely event of their being detected in time. But as Admiral of the Fleet Lord Fisher commented at the start of the century, "The best scale for an experiment is 12 inches to the foot!" And so it was to prove: experience was to show that the elaborate theorizing had been almost entirely wrong.

Let us take a brief look at the characteristics common to almost all homing missiles. They are rocket-propelled, and are accelerated to a very high speed within a few seconds, after which they coast along, gradually losing speed until finally control is lost, when they either explode at the end of their run or fall harmlessly (from the opposing pilot's viewpoint) to the earth. For most of their travel they are too fast to be outrun, so, leaving aside countermeasures for the moment, evasive action is the only recourse.

In the long-gone days when guns were the only effective air-to-air weapon, a standard method of evasion was to dip the nose of the aircraft and accelerate away out of range. Against

Below: A Canadian Armed Forces CF-18 carries eight BL-755 cluster bombs on its four underwing pylons.

missiles, with effective ranges measured in miles, this was futile. The new homing missiles had not ended manoeuvring combat; rather, they had made manoeuvre much more important. Yet the missiles were supposed to be able to out-manoeuvre the fighters. This idea appears to have arisen from the fact that some missiles were advertised as being able to perform 30g turns, whereas fighters were designed to a limit of 7g or a little over.

There are three ways of measuring turning ability. One is in multiples of acceleration, or g; a second is in terms of radius of turn, measured in linear distance; while the third is in rate of turn, expressed in degrees per second. As turning performance is a function of speed, the number of gs that can be pulled is largely irrelevant. The quoted 30g is for a very high velocity only, and as the velocity of the missile decays, so does its ability to manoeuvre. In this it is like an aircraft, but a much more extreme case.

Missile versus aircraft

Major John Boyd's concept of energy manoeuvrability applies to missiles exactly as it does to fighters. A missile travelling at Mach 4 at the tropopause and describing a 30g turn would have a radius of turn of about 14,600ft (4,450m) and a rate of turn slightly exceeding 14deg/sec. By comparison, a fighter flying at its 'corner velocity' – the point where its turning ability is best, or about 400kt (737km/h) – can turn on a radius of only 8,180ft (2,494m) with a mere 2g acceleration, and achieve a turn rate of 16deg/sec at 6g. Naturally, the missile does not necessarily have to match either the turn radius or turn rate of its target, as it can cut the corner, but having to manoeuvre bleeds off energy and reduces its further capability, while at the same time the difficulties of tracking are greatly multiplied. This, of course, applies to a missile coming in from the stern quadrant of the target; from any other direction an energetic manoeuvre may easily take the target outside the missile's flight envelope.

Missiles perform simple tasks best, so the role of the pilot of the target aircraft is to make himself as difficult to hit as possible. Discounting countermeasures for the moment, he does this by generating as much angle-off as he can and hopefully puts himself outside the missile's reach. If he either knows or suspects that the missile tracking him is a heat-seeker, he will attempt to shield his hot exhaust behind the cooler body of his machine.

To summarize, missiles are not yet ten feet tall. Like aeroplanes, they have clearly defined flight performance envelopes, and a pilot under attack will often survive if he can draw it either beyond the limits of the envelope or out toward the boundaries where it will perform less well. The really clever part is knowing when he is under attack, particularly from BVR (beyond visual range) missiles.

While missiles have flight performance envelopes, these are conditioned to a large extent by the relative speeds and aspects of the launching aircraft and its target, so that performance data that is generally available must be qualified. For example, speed is generally stated in terms of Mach number, but many factors affect this, including the velocity of the launching fighter and the altitude at launch. Ideally, the launching fighter should be flying as fast as possible at launch to impart as much of its own velocity, and thus energy, to the missile, in this way maximizing the total energy of the missile at motor burn-out. This in turn will increase range and also kill probability (P_k), especially with BVR weapons.

Launch at low altitude will not only reduce the speed of the launching aircraft, and thus the amount of extra energy that it can impart, but also the maximum velocity of the missile. The air at low level is denser, creating more drag, and the higher pressure on the backchamber of the missile will decrease thrust. Maximum speed data for missiles should therefore be viewed with caution, as, for the same reasons, should flight-time and range.

Still other factors affect speed and range. An obvious example arises in the case of a missile being launched at a target high above, when the missile has to use a considerable portion of its energy in lifting itself to a great height, instead of propelling it as far and as fast as possible, although this is to a small extent offset by the benefits of falling air pressure on the backchamber and reducing drag as the altitude increases and the air gets thinner.

Above: FSD Hornet 7 launches a Sparrow during early armament trials at NAS Patuxent River. Sparrow confers a BVR kill capability.

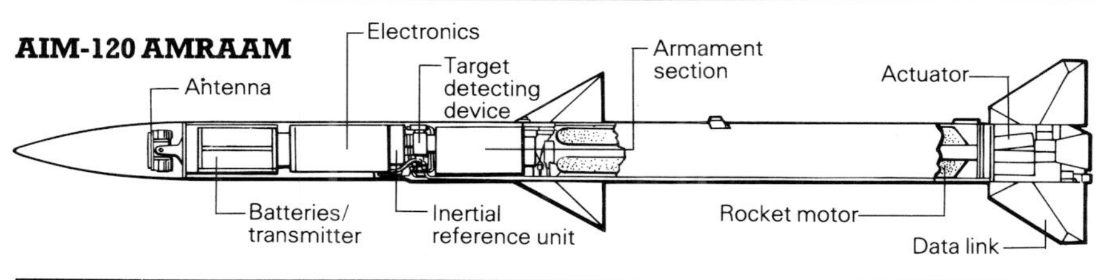

Right: AMRAAM is a launch-and-leave missile using inertial guidance for the midflight phase and active radar terminal homing.

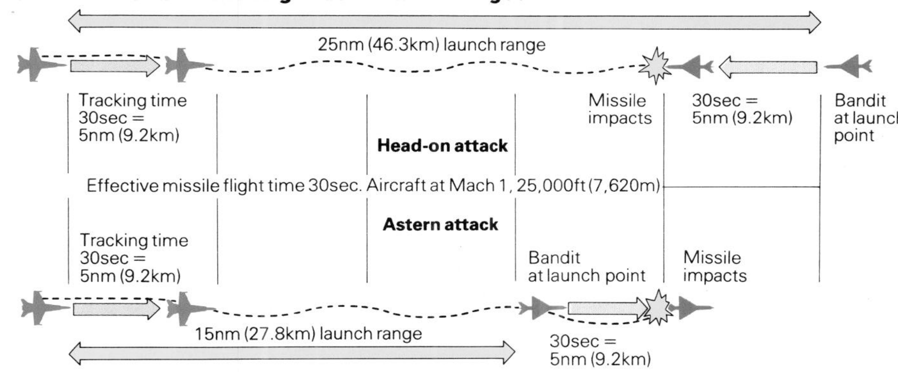

Right: Target speed and relative heading vary an air-to-air missile's launch envelope considerably – by 10nm in this example.

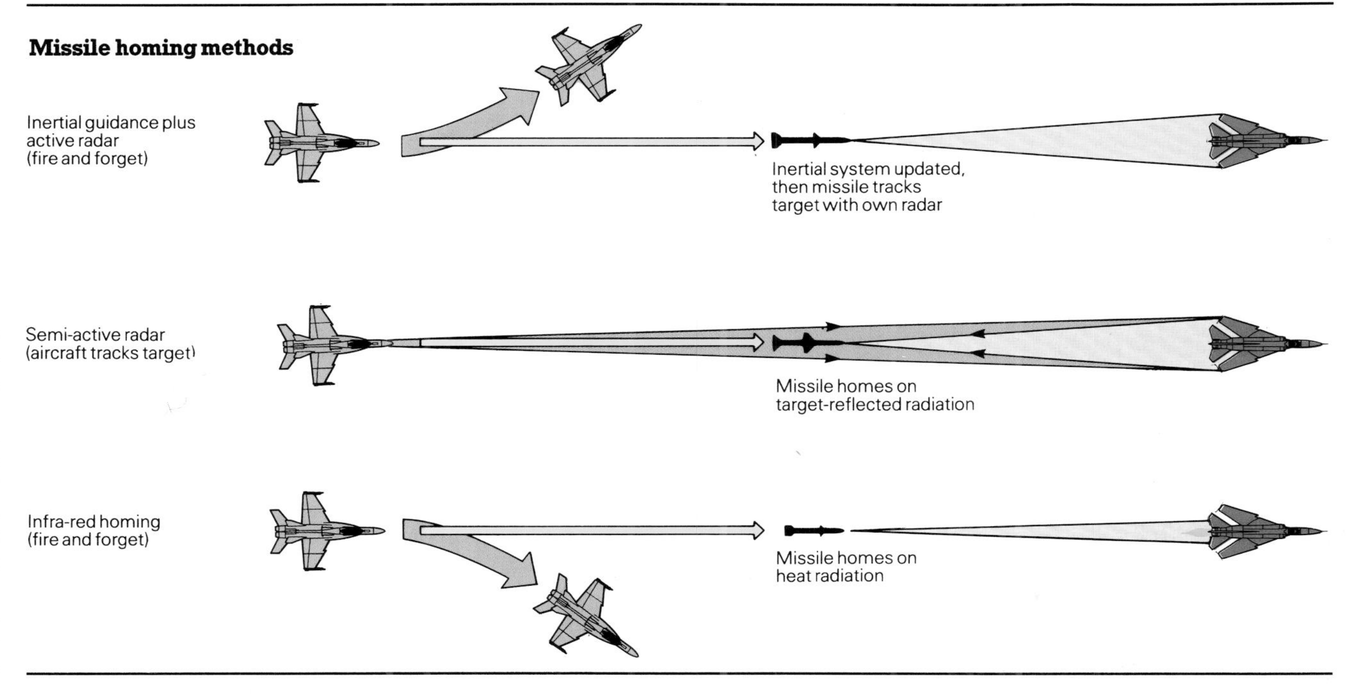

Below: Three types of missile homing: AMRAAM (top), Sparrow (centre) and Sidewinder.

Above: Ears protected against noise and heads helmeted against hard edges, armourers reload a Hornet's M61 Vulcan cannon.

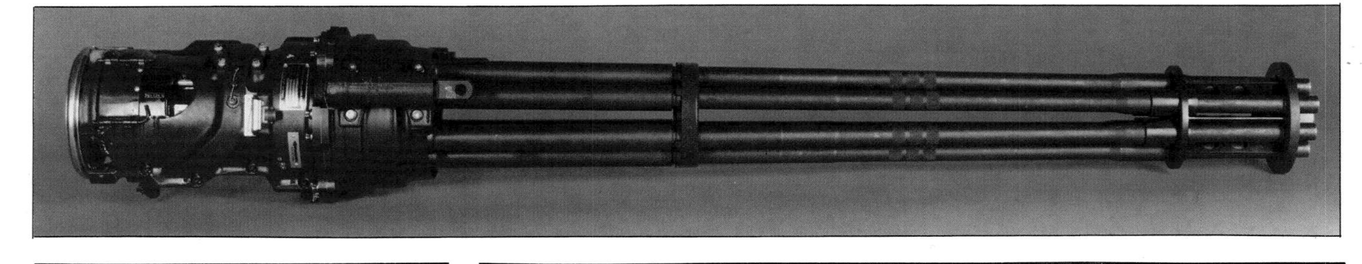

Right: The six barrels of the M61 give a maximum rate of fire of 100 rounds per second and help achieve a high degree of reliability.

Range is generally stated in terms of distance, which is a fixed measurement, but since a missile is launched from a dynamic object at a dynamic target this can be very misleading. If we compare hypothetical cases of head-on and stern-on attacks using a missile with a flight time of 30 seconds and a static range of 20nm (37km), given a target velocity of Mach 1 in both cases, and provided the homing system is up to the task, the missile can be launched against the head-on target while it is still 25nm (46km) away, whereas from astern the missile must be launched at a maximum range of 15nm (28km) to allow it to overhaul the target. The difference is the distance that the target moves during the time of flight. Launch range parameters are thus almost infinitely variable, depending on comparative aircraft velocities, aspects, headings and, of course, relative altitudes.

AIM-7 Sparrow

The primary air-to-air weapon of the Hornet is the AIM-7F Sparrow, and probably in the future the AIM-7M. A semi-active radar homer, conferring the ability to engage targets from beyond visual range, Sparrow originated nearly 40 years ago as Project Hot Shot for the US Navy. Since that time it has been continually up-graded and improved to the extent that it represents one of the biggest missile programmes in history, with a projected grand total approaching 55,000 of all variants built by the time production ends.

M61 hit probability

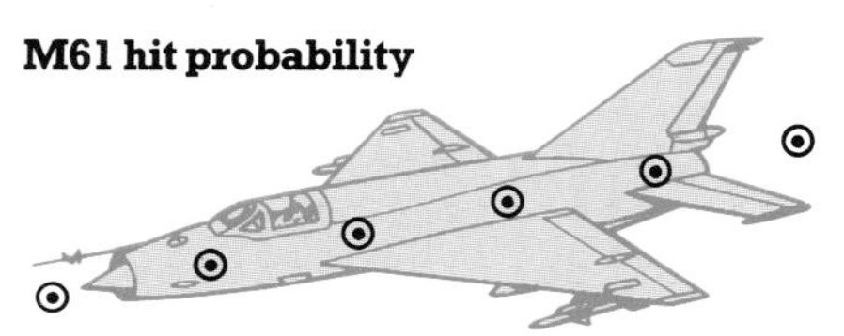

Above: A MiG-21 at 500kt (921km/h) and 90deg angle-off should be hit at least four times by an accurately aimed burst from a Vulcan.

F-18 nose gun and radar installation

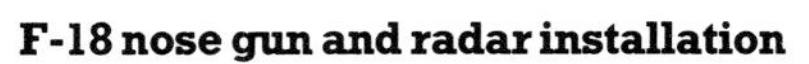

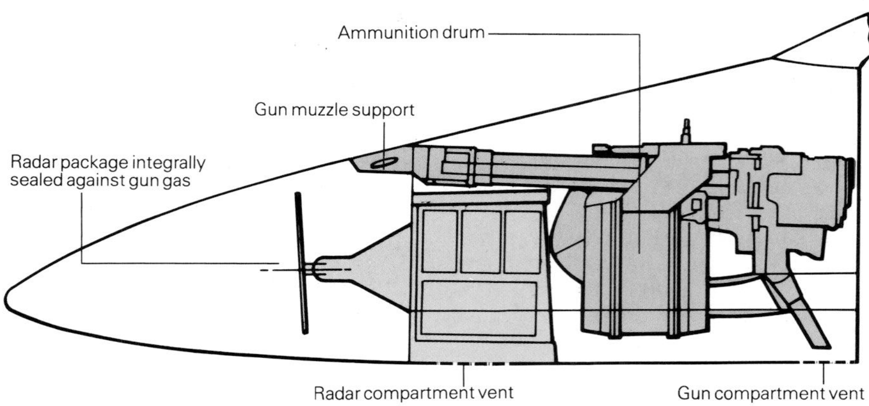

Right: The close proximity of the M61 cannon and the radar equipment in the nose of the Hornet required some very clever design work to damp antenna vibration down to an acceptable 30 g during gun firing.

Sparrow saw extensive service in Vietnam, although its BVR capability was severely curtailed following a couple of unfortunate 'own goals' in the early days. This was not a reflection on the capabilities of Sparrow, merely an indication of the inadequacy of the electronic identification methods of those days. At first, a strict insistence on visual target identification prevailed, although subsequently Sparrow BVR attacks were allowed under strictly controlled conditions. Modern identification systems have improved considerably, although it remains possible that operational restrictions will have to be imposed on BVR attacks in certain situations.

Sparrow offers many operational advantages. While it is hardly sporting, the ability to kill from medium to long range, and certainly from beyond visual distance, has obvious attractions. It also finally nails the old myth about the chivalry of the air, which in truth has never really existed. Air combat is and has always been a matter of creeping up unseen and shooting one's adversary in the back. Victory goes not to the bravest or the strongest, as the old legends would have us believe, but generally to the sneakiest, and few weapons are more sneaky than Sparrow, which comes hurtling in out of nowhere, fired by an unseen assailant.

Used in conjunction with the parent fighter's pulse-Doppler radar, which illuminates the target with the radar impulses whose echoes the Sparrow homes on, the missile is at its best against a head-on target with a high closing rate. As was sometimes the case in Vietnam, opposing fighters are forced to run the Sparrow gauntlet before they can close the range to a point where they can fight back. A classic instance of this took place near Yen Bai on the morning of May 10, 1972, when the four Phantoms of Oyster Flight took out two of four intercepting MiG-21s before closing to visual distance.

Sparrow has come in for heavy criticism in recent years because its SARH system demands that the parent fighter

illuminate the target with its radar during the entire homing phase. A second, and far more valid criticism is that it forces the pilot to concentrate on one target to the exclusion of all others. But to return to the first point, the launch aircraft, having fired a Sparrow, has to fly a more or less head-on course towards the target to illuminate it. The supposedly inevitable consequence is that the target makes a visual sighting just before the Sparrow blows it from the sky, and fires a launch-and-leave missile of its own, the destruction of the target being followed in short order by the destruction of its attacker.

Even with the advantage of numerical superiority, swapping one for one is not a good way to fight a war; it certainly does nothing for a pilot's morale. But while the argument has a great deal of merit, it seems unreasonable to assume that a one-for-one swap will be the outcome in a majority of cases, even though it must be admitted that the risk does exist. There must be many cases when the target will be splashed without seeing its assailant, particularly when there is a great height disparity between the two, and especially when the Sparrow has been launched from a lower level. A beam or front quarter attack would also reduce the possibility considerably.

In this connection, the Hornet is a fairly small aeroplane with smokeless engines and, very importantly, small engine intakes tucked away beneath the wings. It is a very difficult aeroplane to spot from head-on. And in mediocre weather visual acquisition will be even more difficult. Much of the case against Sparrow arose from the AIMVAL/ACEVAL series of exercises, which were held in Nevada, where the climate and visibility are near perfect for much of the time.

Nevertheless, there can be no doubt that having to illuminate the target with radar for relatively long periods is tactically undesirable, the illuminating fighter being visible and predictable for far too long. Using the radar for protracted periods is like lighting a huge electronic beacon in the sky saying, "This is where I am, this is who I am, and this is what I am doing". There can be no doubt that a BVR launch-and-leave weapon would be preferable to Sparrow, but to look at the positive side, Sparrow does confer genuine BVR capability.

Psychological impact

Another positive aspect, which unfortunately cannot be quantified, arises from the emissions that accompany a Sparrow launch. The chances are that the RWR of the target has warned the pilot that a Sparrow-type radar is locked on to him, as it can detect when a hostile radar changes from search to attack mode, but he has no way of knowing whether a missile has been launched or not.

There is an element of psychological warfare here: the pilot of the target (or group of pilots in close formation) will feel threatened by the radar emissions, which may tend to lower their mission effectiveness. While the launch-and-leave weapon is tactically preferable as a weapon, it can have no such effect, although one psychological effect common to all BVR weapons is the bolt from the blue, when one of the formation is suddenly blown away without any prior warning. Even the sight of a big Sparrow steaming past in the event of a near miss will hardly be reassuring.

The Sparrow is a big missile. It is 8in (20cm) in diameter and 12ft (3.66m) long, with a 3ft 4in (1.02m) span. Maximum speed is about Mach 4, and the stated range of the AIM-7F is 62 miles (100km). One telling comment on the missile's size was provided by a paper study done a few years ago to assess the effect of equipping the F-5E to carry it, when the general effect was described as "like putting an anchor on the airplane". The Hornet, however, is large enough and powerful enough to carry it with no problems.

Above: FSD Hornet 8 during gun firing trials in April 1980. Unusually for a modern fighter, the cannon is located just above the aircraft centreline. This is as close to the ideal position as the radar allows.

Below: A Hornet of VMFA-314 Black Knights lets fly with a Sidewinder from the starboard wingtip rail. Previously equipped with F-4Ns, the Black Knights converted to the Hornet in March 1984 and are assigned to CVW-13 on USS *Coral Sea*.

Left: **FSD Hornet 7 configured for the attack mission, with three Paveway laser-guided bombs and an AGM-84 Harpoon on pylons, Sidewinders on the wingtips and fuselage FLIR and LST/SCAM pods.**

been manufactured in greater numbers than any other Western missile, with production currently approaching the 130,000 mark.

Sidewinder is essentially a visual distance weapon. The seeker head developed for the L variant uses Argon-cooled Indium Antimonide, which is extremely sensitive to IR emissions, and has an all-aspect capability. An aircraft in flight is warmed by friction heating due to the air passing over its skin. This is more pronounced where the local airflow is accelerated, such as the nose, or the leading edges of the flying surfaces, and the seeker head is sufficiently sensitive to detect this heating against the cold ambient background of the sky.

Clever filters are incorporated to ensure that the missile does not home on the sun or other IR source, although in the Gulf of Sidra incident in August 1981, when Libyan Air Force Su-22s clashed with US Navy F-14s, one of the Tomcat

First introduced in 1977, the AIM-7F Sparrow has Raytheon solid-state guidance with a conical scan seeker head. Propulsion is by the Hercules Mk 58 solid fuel motor, and the warhead consists of 88lb (40kg) of high explosive, contained in a drum made from a continuous stainless steel rod which shatters into about 2,600 fragments on detonation. With these tiny, high-velocity fragments flying about, the probability of lethal damage being caused to the target is very high. Detonation is triggered either by impact or by proximity fuze.

The latest Sparrow variant is the AIM-7M, which is fitted with an inverse monopulse seeker and a digital signal processor, which should give improved look-down and ECCM capability. Also featured are a new autopilot and a new fuze. Hornet will carry the AIM-7F at first, but is expected to convert to the AIM-7M at some future date.

To counteract the shortcomings of Sparrow, the AIM-120 AMRAAM was developed by Hughes. A launch-and-leave missile, it initially flies towards its target using inertial mid-course guidance, then activates its own X-band radar seeker in the nose for the terminal phase of flight. As there is no necessity for the parent fighter to illuminate the target, this missile will give the Hornet a multi-shot capability, engaging more than one target simultaneously rather than having the radar concentrate on one target to the exclusion of all else. AMRAAM has successfully demonstrated a look-down capability in test firings, but little firm information is available. Its overall dimensions are similar to those of Sparrow, but at 326lb (148kg), it is much lighter. Maximum speed is estimated to be about Mach 4, and its range is believed to exceed 30nm (55km).

AIM-9 Sidewinder

The AIM-9L Sidewinder carried by the Hornet is almost certainly one of the last variants of a very long line dating back to 1949. It started life as a simple and very cheap (about $2,500 at the time) missile using an infra-red (heat) homing system. From dead astern, aimed at the hot jet efflux of a non-manoeuvring target, it was very reliable, but targets are rarely cooperative. As the years passed, it was developed into a much more capable weapon with greatly increased range and tracking ability, but at the penalty of vastly increased cost.

Above: FSD Hornet 4, the structural test prototype, photographed in August 1981 carrying four Mk 84 2,000lb (907kg) on wing pylons. Blue and white Sidewinders are carried for attitude identification.

Right: Close-up of the underwing pylons, the nearer mounting a pair of Mk 82 slicks on a twin store carrier. Underwing stores carriage is a cumbersome business.

All versions before the L were stern attack weapons, although the J can under some circumstances acquire targets from other angles, and it is generally considered that if the pilot of the target aircraft knows that his opponent is behind him, and if he has sufficient energy manoeuvrability, he can deny a valid Sidewinder shot. Sidewinder has pilots reported that he waited until the target cleared the sun before launching a Sidewinder, which seems to imply a certain lack of faith on his part – or, of course, total professionalism! Sidewinder has many advantages over Sparrow, not the least of which being that it is a launch-and-leave weapon. It is small, with a diameter of only 5in (12.5cm) and a length of 112.2in (2.85m), and it weighs only 188lb (85kg). It can be fitted to almost any aircraft with ease, needing little more than a launch rack or rail, a few wires, and earphones. The wingtip mounting rail on the Hornet has the advantage of increasing the effective aspect ratio of the wing, which reduces its lift-dependent drag. This confers better cruise performance and sustained manoeuvrability compared with underwing racks.

The Hornet's wings fold to reduce storage and deck space required. When it is being armed before a mission, it is obviously an advantage to be able to load the Sidewinders onto the wingtip rails with the wings in the folded position. This is possible, although the safety aspects are reported to give ordnance officers pink fits! The main disadvantage of Sidewinder is that its detection range drops dramatically in cloud or heavy rain; in extreme conditions it becomes unusable.

Sidewinders in combat

Sidewinder has an operational history dating back to September 1958, when Sabres of the Chinese Nationalist Air Force, operating from Taiwan, clashed with communist Chinese MiGs. Four MiGs were claimed to have fallen to Sidewinders in what was the very first operational use of any air-to-air homing missile. In the Vietnam war early-model Sidewinders were widely used, achieving a kill ratio of about 15 per cent. This was considerably better than the kill ratio achieved by the Sparrow, which varied between 8 and 10 per cent. Sidewinder is an inherently more accurate weapon than Sparrow, due to its ability to 'see' a heat source more clearly than Sparrow can sense the fuzzy, wandering epicentre of a target provided by reflected radar emissions. This is one reason for the Sidewinder's better kill ratio, but it should be remembered that it was a better and more reliable short-range weapon than Sparrow and was consequently used on the easier targets.

The AIM-9L received limited use by Sea Harriers in the South Atlantic in 1982. There were 23 missile engagements in which a total of 26 missiles were fired, causing the destruction of 19 Argentinian aircraft. On three occasions two missiles were fired at the same target. This gives a kill ratio of 73 per cent. While this record is very impressive it should be noted that the all-aspect capability of the AIM-9L was not really tested. Most missile firings were from astern and from fairly low angles-off and rarely were the targets manoeuvring energetically. An Argentinian pilot later commented that missiles could be out-manoeuvred if they were seen coming in time; he was referring to surface to air missiles during low level attacks but his remark has a certain validity. On one occasion, two Sidewinders were launched without success at an Argentinian Canberra manoeuvring hard at low level over the sea, while on others the missile succeeded in tracking and hitting evading fast jets.

Right: A Hornet of VFA-113 Stingers en route to the Leach Lake range in California (top). After releasing two Mk 82 slicks in a shallow dive (centre), the aircraft banks away (bottom), showing the empty racks.

Above: The BL-755 cluster bomb unit is not a new weapon, but it still had to be cleared for use by the Hornet. This Canadian CF-18 carries eight, with cameras attached in five positions to record the drop.

Sidewinder is very easy for the pilot to use. With the seeker head uncaged, i.e. live, the missile announces that it is looking at something warm with a noise variously described as a 'growl' or a 'rattle' in the pilot's earphones, which increases in intensity as the target indications improve. The missile is then locked on to the target it is looking at and a squeeze of the trigger sends it on its way. AIM-9L has a stated range of 11 miles (18km) and a speed of about Mach 2.5, with a flight time of one minute. Its warhead is of the annular blast fragmentation type, with 22lb (10kg) of high explosive, and it carries fuzes for detonation of both impact and proximity types.

Gun armament

In common with all modern fighters, the Hornet carries a gun. Once upon a time, when the new wonder missiles were emerging, guns were thought to be superfluous except for ground strafing. But that is another story, and not everybody believed the theory anyway. Soviet fighters were almost without exception armed with guns; notable exceptions to the missiles-only trend in the West were Dassault with the Mirage III and, of course, Northrop with their F-5.

One further factor, which has a certain relevance to the Hornet, was that Western fighters of the period were being made far more capable than hitherto, able to fly their missions at night and in adverse weather conditions. This, plus the illumination requirements of the SARH missiles, resulted in the nose of the fighter being almost entirely occupied by a large radar set, while wing mountings, so widely used in World War II, apart from being poor places to mount guns, were hardly practicable, as the demands of Vmax had made them too thin. It was consequently difficult to find a suitable location which would also avoid gun exhaust gases being ingested by the engine with undesirable results.

Combat experience in the skies over North Vietnam soon had the American fighter pilots screaming for guns. With visual identification requirements reducing Sparrow effectiveness considerably, and a cruising speed of about 480kt (884km/h), the missile-only Phantoms often found themselves embroiled in 'knife-range' combats with the North Vietnamese MiGs. On May 4, 1967, Major Bill Lafever, then a young Lieutenant, was flying as backseater to the famous commander of the 8th TFW, Colonel Robin Olds. After a hectic fight in which four Sparrows and three Sidewinders (the other suffered a malfunction) were fired and a MiG-21 shot down, Olds led his flight straight over Hoa Loc airfield. There were about five MiG-17s in the circuit and Olds made several dummy passes on them at extremely close range. Without a gun, he could do little else.

MiG kill statistics

More than ten years afterward, Bill Lafever's comments on gunless fighters were still vividly descriptive, although hardly fitting for these pages. The two months immediately prior to the encounter just described had thrown up a relevant statistic. During this period 10 MiG-17s were downed, all by gunfire from the huge and unwieldy F-105 Thunderchief fighter-bombers, while their missile-only Phantom escorts could claim but two MiG-21s.

The M61A1 Vulcan cannon was the gun chosen to arm the Hornet, with 570 rounds of 20mm ammunition allowing about six seconds of firing at its maximum rate of 100rds/sec. The Vulcan utilizes six rotating barrels on the Gatling principle and is a fairly large weapon, so that locating it in a smallish aeroplane presented problems. The best place for any aircraft gun is on the centreline of the aircraft, where the recoil causes no asymmetric loading, but an immediate problem arises from the fact that this position is invariably occupied by the radar. The priority then becomes finding the best possible compromise position.

The same problem had been encountered on both the F-15 and F-16. On the F-15, the gun has been mounted in the starboard wing root, while on the F-16 it is accommodated in the port side of the fuselage, and both aircraft have the magazine in the centre fuselage behind the cockpit. In both cases it was considered preferable to accept the disadvantages of asymmetric mounting – even to the extent of automatically calling up rudder to counteract the yawing moment generated when the gun was fired – in order to keep the tremendous vibration caused when firing the gun away from the sensitive radar. In the Hornet, the designers took the bull by the horns and placed the gun in the nose, just above the centreline, in close proximity to the radar. The gun and its magazine were pallet-mounted to give quick and easy access for changing.

The vibration problem was then

Above: One of the principal anti-ship weapons carried by the Hornet, the AGM-65F variant of Maverick uses imaging infra-red homing.

tackled. Ground firing trials showed that the vibration level at the top of the planar antenna could reach a totally unacceptable 400g. Hughes overcame this problem by putting four anti-vibration mounts on to each bulkhead at the back and front of the radar. This allowed the radar assemblies to 'float' between the bulkheads and damped the vibration down to about 30g, which, although it sounds a lot, was reckoned acceptable.

Only the position of the muzzle, on top of the nose and immediately ahead of the pilot, appears suspect. When the gun is fired at night, the muzzle flash would seem to be in the perfect position to interfere with the pilot's night vision. As recounted earlier, a night-firing trial was held, after which it was reported that no difficulties were encountered in tracking a flame float at night, but a flame float is an entirely different proposition to an unlit air or ground target.

The General Electric M61A1 Vulcan cannon has been the standard USAF aircraft gun for many years, and is a most reliable weapon, with an average stoppage rate of just one round in 10,000. One reason for its reliability is the fact that it is externally powered, and the feed does not depend on the gun's own action, unlike the previous generation of gas-operated revolver cannon such as the M39. The six rifled barrels rotate anti-clockwise (looking in the direction of fire), and the fact that there are six barrels reduces wear and heat dissipation, giving longer weapon life.

The tremendous rate of fire requires a linkless feed; it also gives a very high hit probability against a fast-moving, high-angle-off target. M50 series ammunition is used, with a muzzle velocity of 3,400ft/sec (1,036m/sec). With a firing rate of 6,000 rounds per minute, the shells are spaced 34ft (10.36m) apart on leaving the muzzle, and this distance reduces during the time of flight. A 44ft (13.46m) long MiG-21 Fishbed at 90deg angle-off and with a velocity of 500kt (921km/h) travels its own length in 0.052 seconds. If the aim of the Vulcan is accurate and the target motion is in plane, the Fishbed is going to be hit at least four times.

A criticism of the Vulcan heard from some quarters is that it takes time to wind up to the full rate of fire. In fact, it takes all of 0.3 seconds, and a further half-second to wind down again. The argument goes that the first placement of the gun aiming dot, or pipper, on the target, is generally the best; after that it tends to wander off. Because of this – the criticism stems from a paper written a few years ago by defence analyst Pierre Sprey entitled *First Rounds Count* – the time taken for the Vulcan to wind up to full speed is considered a disadvantage, as it is still gathering speed at a time when the aim is at its truest.

Below: Infra-red images produced by the AGM-65F's IIR seeker and displayed on the cockpit CRT: the destroyer USS *Bagley* at the limit of the pilot's visual range (upper) and at the terminal homing stage.

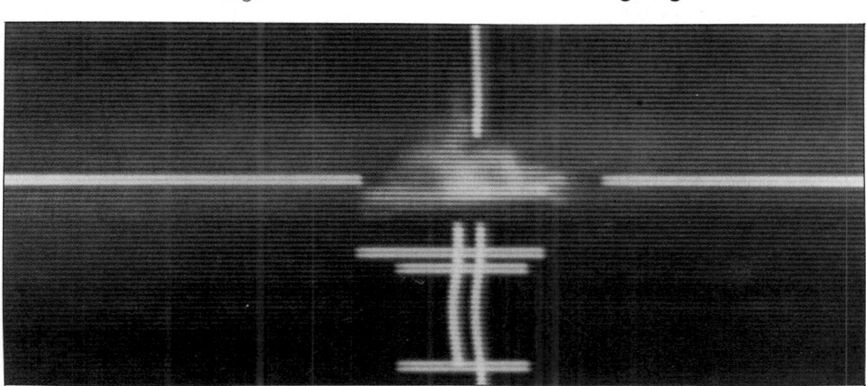

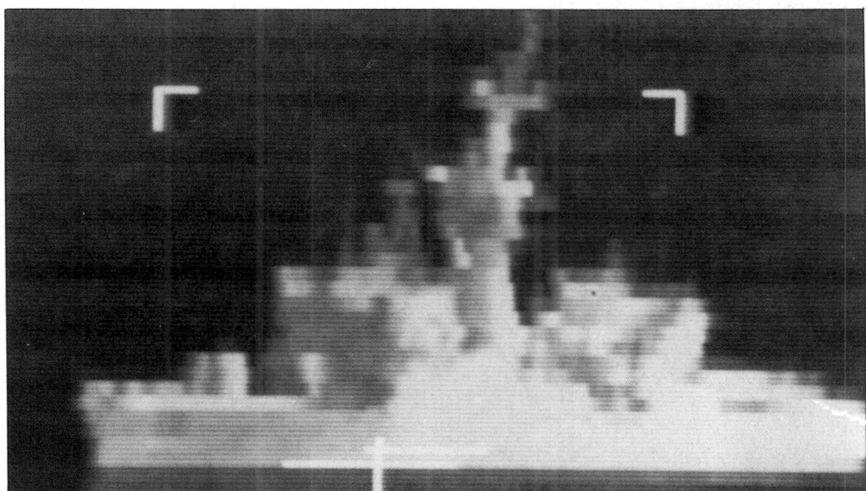

Below: Another weapon test carried out by the CAF Aerospace Engineering Test Establishment, this time involving unguided rockets.

The answer to this is fairly simple. Gas-operated revolver cannon such as the M39 do hit their full rate of fire instantly. The trouble is that the full rate of fire for an M39 is only 1,500rds/min, so it would take four of them to exceed the output of the M61 in the first half-second of firing. Fitting one gun, albeit a bulky one, into a Hornet was a difficult task. Where would four M39s go?

The aircraft gun can be summarized as an essential close-range weapon of great reliability, with a snap-shot capability unequalled by any missile yet built. It also increases the number of potential on-board kills considerably: each missile can be used but once, whereas the gun is a repeating weapon. Short on range and of limited application it may be, but an adversary ignores it at his peril. Finally, the only effective countermeasure against it is manoeuvre. The radar may be solid with jamming and decoys may render heat missiles useless, but the gun-armed fighter will always possess the means to shoot an opponent from the sky.

Ground-attack weapons

The load-carrying capacity of the Hornet is stated to be a maximum of 17,000lb (7,700kg), though it seems that this weight is unlikely to be carried operationally. The two wingtip Sidewinders will be carried on the attack mission to give a self-defence capability, but the Sparrows will be supplanted by the FLIR and LST/SCAM pods.

If long range or endurance is required, the centreline and the two inner wing hardpoints are wet and can be used to carry jettisonable tanks which add 6,435lb (2,920kg) of fuel. While multiple ejection racks (MERs) on the four wing hardpoints and a triple ejection rack (TER) on the centreline allow a total of 19 Mk 82 bombs of 500lb (227kg) nominal weight each to be carried, this is not a typical load. The drag penalty of loaded MERs and TERs is very high, and against strongly defended targets it is possible that only twin store vertical ejection racks (VER-2s) will be carried. A typical load would therefore be four Mk 83s and six Mk 82 low drag bombs (slicks), carried in pairs.

The drag penalty of MERs and TERs arises not only from the racks and weapons considered individually, but also from the interference drag they cause in combination. In flight, the airflow accelerates past the bombs and their racks, reaching a maximum just ahead of their maximum width. Consequently, clusters of bombs grouped close together on an MER impinge upon each other's and the rack's accelerated airflow, and interference drag results. This affects not only aircraft performance but also bombing accuracy, as drag-induced forces tend to impart pitching and yawing moments to the weapons at the moment of release, especially with current attack speeds approaching 600kt (1,110km/h).

Operational handling of the bombs is simple. The ground crew load the weapons and make the fuze code settings. The pilot then checks the loading on the MMD, and sets up to three delivery programs on the attack computer. Even these are not immutable, as both the programs and the weapons' status can be changed during flight.

The range of weapons that can be carried by the Hornet is fairly comprehensive. Apart from the Mk 82, 83, and 84 slicks, laser-guided versions of all these weapons can be carried. The principle of laser-guided bombs (LGBs) is fairly simple: the target is marked by a laser designator, either from the ground or from another aircraft, which produces a funnel of reflected laser light over the target, rather like the basket in basketball. The trick is for the attacking aircraft to deliver the bomb into the top of the funnel, whereupon it will glide down on to the target.

The LGB can be delivered into the top of the funnel in a variety of ways. In a heavily defended area, the dive-toss method is preferred: the attacking aircraft approaches low and fast, then at a pre-calculated point and speed, pulls up and launches the bomb in a high trajectory towards the top of the funnel.

Also in the inventory are Mk 82 retarded bombs, which allow the attacker to get clear of the explosion of a bomb dropped from very low level. Various

US Navy F-18 ordnance loads

Weapon	Armament Station 1	2	3	4	5	6	7	8	9
Air-to-air missiles									
AIM-9G/H/L Sidewinder	1	2						2	1
AIM-7F Sparrow		1		1		1		1	
Air-to-surface missiles									
AGM-65E/F Maverick		1	1				1	1	
AGM-88A Harm		1	1				1	1	
Conventional weapons									
Mk 82 LD/HD		2	2		2		2	2	
Mk 82 LGB		1	1				1	1	
Mk 83 LD		2	2		1		2	2	
Mk 83 LGB		1	1				1	1	
Mk 84 LD		1	1				1	1	
Mk 84 LGB		1						1	
Mk 20 or CBU-59/B Rockeye		2	2		2		2	2	
BLU-95 FAE-II (fuel-air explosive)		2	2		2		2	2	
AGM-62 Walleye		1						1	
Walleye data link pod					1				
Practice bombs									
Mk 76/Mk 106 dispensers		1	1		1		1	1	
BDU-12/20		1	1				1	1	
BDU-36		1						1	
Rocket launchers									
LAU-10D/A, -61A/A or -68B/A		2	2				2	2	
Special weapons									
B57 or B61		1						1	

Above left: Defence suppression is a task that an air arm ignores at its peril. Here, the first firing of an AGM-88A Harm by an F-18 is carried out by FSD Hornet TF1 over Wallops Island test range in October 1983.

Left: The cameras carried during separation trials show up clearly on wingtips, in Sparrow positions and under the rear fuselage as an inert B61 'special weapon', designated BDU-12/20, falls clear of FSD Hornet 7.

rocket launchers can be fitted – LAU-10/D, -61A, and 68B/A – while among the more exotic weapons are the fuel/air explosive BLU-95/B and the CBU-59B Rockeye cluster bomb, which covers the target area with a pattern of bomblets, effective against soft vehicles, personnel, and parked aircraft.

Among the 'smart' weapons carried is the AGM-62 Walleye glide bomb, an electro-optically guided weapon with a data link for extended range. Walleye is released in the general direction of the target then, via the television camera in the nose of the weapon, the pilot acquires the target on his monitor screen. He then focusses the camera on the target and locks it on. A powered weapon of similar concept is the AGM-65A/B Maverick, which has a range of 14 miles (22km), although other versions of this weapon use imaging infra-red (IIR) to provide the target picture, or laser seekers; respective designations are AGM-65D and AGM-65C/E.

Against shipping targets, AGM-84 Harpoon is used. Harpoon is a long-range – 68 miles (109km) – missile: steering commands are programmed into the strapdown inertial system before launch, and height above the sea is controlled by a radar altimeter, while propulsion is by a small turbojet which gives it a high subsonic speed at very low altitude. Skimming in low over the water, at a predetermined point it switches on an active radar seeker which searches, then automatically locks on to a target. During the final phase of the approach it pulls itself up, then dives on to the target from above. As it is a launch-and-leave missile, the Hornet will at no time have to make a close approach to a hostile fleet in order to carry out a Harpoon attack. Supersonic and IIR versions of Harpoon are under consideration.

AGM-88A Harm

A totally different type of weapon is AGM-88A Harm (High Speed Anti-Radiation Missile). Harm is a radiation-seeking missile with a range exceeding 11 miles (18km), and superior detection capabilities and higher speed than either of its predecessors, Shrike and Standard. Its function is to seek and destroy hostile ground radars, and it operates in three modes: self-protect, target of opportunity and prebriefed.

In the self-protect mode, the Hornet's RWR will detect threat emissions and the mission computer will assess threat priorities, prepare a program and feed the necessary data to the missile, all in a matter of milliseconds, whereupon Harm can be launched and will home on the emissions selected. In the target of opportunity mode, the missile seeker operates automatically using methods which are carefully concealed under a blanket of woolly terminology. In the prebriefed mode, Harm is used in a rather speculative fashion. It is launched from within range and in the direction of known emitters (radars): if one comes on the air and starts to radiate, then Harm will immediately home on to it, whereas if nothing happens, Harm explodes in mid-air at the end of its run.

With this arsenal of weapons at its disposal, Hornet is a very potent fighting machine indeed. Whether it operates as a fighter or as an attack aircraft, it has the systems and the weaponry to successfully tackle a broad spectrum of threats, tasks and targets.

F/A-18 Hornet stores options

1 AIM-9L Sidewinder AAM
2 AIM-9J Sidewinder AAM
3 AGM-65 Maverick ASM
4 AGM-62 Walleye ASM
5 AGM-109 Harpoon anti-ship missile
6 Drop tank, 315US gall (262 Imp gall, 1,192lit)
7 B57 tactical thermonuclear bomb
8 Durandal anti-runway weapon
9 SUU-20 practice dispenser
10 ASQ-173 laser spot tracker/strike camera (LST/SCAM) pod
11 AIM-7 Sparrow AAM
12 AGM-88A Harm anti-radar missile
13 Gun port
14 M61A1 Vulcan 20mm cannon with 570-round ammunition drum
15 GBU-10E/B Paveway II Mk 84 2,000lb (907kg) laser-guided bomb (LGB)
16 AAS-38 forward-looking infra-red (FLIR) pod
17 Mk 84 2,000lb (907kg) low-drag (LD) general-purpose bomb
18 Three Mk 82 500lb (227kg) low-drag (LD) 'slick' general-purpose bombs on TER (triple ejection rack)
19 Mk 82 high-drag (HD) 'Snakeye' retarded general-purpose bomb
20 M117 750lb (340kg) general-purpose bomb
21 Stores carrier
22 Data link container for Walleye guidance or flight test monitoring
23 CBU-59/B or Mk 20 Rockeye antitank cluster bomb
24 Two Mk 83 1,000lb (454kg) low-drag (LD) general-purpose bombs
25 LAU-61A/A rocket pod
26 LAU-68B/A rocket pod

Above: Some idea of the wide variety of stores that the Hornet can carry is given here, though the list is by no means comprehensive. The operational use of TERs is restricted because of the high degree of drag they induce in flight.

Hornets in service

Above: Two F-18As and two TF-18As of the Rough Raiders in a neat formation during the deployment to MCAS Yuma in January 1982.

On June 30, 1981, Vice Admiral Wesley L. McDonald, US Navy, gave his verdict on the Hornet's value to the service: "The versatility of the F/A-18 to effectively perform both the fighter and the attack missions provides the battle group commander with options never before available. When in a defensive posture, the Hornet will counter either air or surface threat. Offensively, it will provide both fighter escort and a survivable ordnance delivery vehicle with finite accuracy. This force multiplication effect is not available with any other aircraft in the world . . . All indications are that the Navy/Marine team has a superb machine in which to move forward into the future."

Experimental squadrons VX-4 and VX-5 were the first USN units to fly the Hornet, but the first true Hornet squadron was VFA-125 – the Rough Raiders – which was commissioned at NAS Lemoore, California, on November 13, 1980. VFA-125 is a fleet readiness squadron, charged with training both ground and air crew for the operational units. As such, it will have at its peak a far greater complement than a normal squadron, with a total strength of 60 Hornets, a high proportion of them two-seaters, 75 officers, including 30 pilots, and about 600 enlisted men. It will convert operational squadrons to the Hornet at a rate of four per year. A second fleet readiness squadron, VFA-106 Gladiators, was commissioned at NAS Cecil Field, Florida, in April 1984, while a third is planned to be activated at MCAS Yuma, Arizona, probably in late 1986.

The Rough Raiders are a unique mixed Navy and Marine Corps outfit. Their first commanding officer was Capt. James W. Partington, USN, and the executive officer was Lt. Col. Gary R. VanGysel, USMC. Their backgrounds were, as one might expect with a dual-role aircraft like the Hornet, dissimilar. Captain Partington has extensive attack experience flying Skyhawks and Corsairs, while Lt. Col. VanGysel is a very experienced Phantom driver. At first, the squadron build-up was slow. The first Hornet arrived on February 19, 1981, and was followed by two more aircraft from the pilot production batch. The first full-scale production Hornet did not arrive until September, and by the end of the year, the Rough Raiders had eight Hornets.

Approximately 150 of the Hornets on order will be two-seat TF/A-18As, about one in every nine of the total order, and following the precedent of the F-15 and F-16, they will probably be redesignated F/A-18B at some future date. The two-holers are fully combat-capable, although carrying less fuel, they are slightly short on range compared to the single seater. They feature an extended canopy, with the rear seat set 6in (15cm) higher than the front seat to give the guy in the back good visibility, and the rear cockpit is identical to the front except that it does not have a HUD. The second seat displaces a fuel tank, and the TF/A-18A carries 600lb (272kg) less fuel internally. Any performance differences between the two are marginal.

By August 1981 the VFA-125 pilots had reached the stage in their training where they were ready for Air Combat Manoeuvring (ACM) experience, despite the fact that at this point only three Hornets had been assigned to the squadron, and these had to be shared between 16 pilots. ACM is an unproductive exercise when carried out between fighters of the same type as the only difference is that of pilot quality, so adversary aircraft were provided in the form of an A-4 Skyhawk of VA-127 (the Cylons), from NAS Lemoore, and an F-5E Tiger II from the US Navy Fighter Weapons School (usually known as 'Top Gun') at NAS Miramar, near San Diego.

An adversary aircraft is a type selected for its performance similarities to known or likely 'threat' aircraft. The Skyhawk represents the MiG-17 while

Below: A Hornet of VFA-125 takes the wire aboard USS *Constellation* during the Rough Raiders' first carrier qualifications in October 1982.

the F-5E doubles for the MiG-21. In most cases, adversary aircraft wear Warsaw Pact camouflage, with Soviet-style 'buzz numbers' on the nose. Adversary pilots are trained in Soviet techniques and tactics, and with their specially chosen aircraft simulate the most realistic threat possible for training purposes.

The four-day exercise consisted entirely of one versus one encounters. The adversary pilots, both very experienced men, were as keen as anyone to see what the Hornet could really do, especially after all the controversy that had surrounded it. They were most impressed, Major George Stuart of the USMC describing the Hornet as being as capable as any aircraft in the inventory, while Lt. John C. Forrester, USN, felt afterwards that there was no comparison between the Hornet and the two adversary types. He was also impressed by the fact that the Rough Raiders, although inexperienced on the type, had reached such a high level of competence, the sign of an easy aeroplane to fly.

The next stage was a ten-day deployment to MCAS Yuma with five Hornets in November 1981, followed by an extended deployment, also to Yuma, by nine Hornets, lasting from January 5-27, 1982, to complete ACM training for the initial batch of 16 pilots assigned to VFA-125, and to finalize the ACM syllabus for the squadrons to be trained on the Hornet. The advantage of Yuma is that it has an electronically instrumented range able to track up to eight aircraft at a time, while simulating and assessing missile firings and recording the proceedings on video tape. The entire mission is then replayed to the pilots on their return. Skilled debriefers beware! This elaborate system, produced by Cubic Corporation's Defense Systems Division in San Diego, rejoices in the acronym of TACTS/ACMI (Tactical Aircrew Combat Training System/Air Combat Manoeuvre Instrumentation). Electronic pods are attached to the participating fighters: these emit signals which are received by a network of solar-powered antennas on the ground for transmission back through a series of micro-wave relay stations to a mobile recording centre. The combats can then be reconstructed and evaluated.

The bulk of the opposition for the second deployment was provided by Skyhawks of the Cylons, although Top Gun Skyhawks, Canadian F-5s from No. 433 Squadron, and Tomcats from VF-51 and VX-4 also participated. The anti-Tomcat sorties were flown one versus one, while the others were generally multi-bogey encounters at odds of up to three to one against the Hornet. The multi-bogey engagements pressurized the Rough Raiders into using their systems capability to the full, whereas in one versus one encounters there is always a tendency to close quickly to visual range, then stay visual.

Naturally, the Hornet did not win every encounter – there is no aircraft/pilot combination in the world capable of pulling that off – but the results achieved left the Rough Raiders full of enthusiasm for their mount. Navy Lt. Phil Scher of the Cylons hitched a ride in the back seat of a Hornet during one exercise. An experienced adversary pilot who had previously flown against the Tomcat, Eagle, and Fighting Falcon, he commented afterwards: "From close up, I can honestly say that the F/A-18 is magic. I don't think that there are many airplanes, if any, that are capable of physically beating the Hornet in the air."

Shoot-out at Yuma

Questions that are frequently posed are, how good is the Hornet in the air-to-air arena, and how would it shape up against the F-16? The following account of an engagement that took place during the second Yuma deployment in part answers the first question.

Through Telegraph Pass at 420kt (774km/h), two F/A-18 Hornet jet fighters enter the range, level at 'Angels one-five', and scan the early morning sky for the enemy. Hornet One's radar locks on to a target. He transmits:

"Contact . . . a single . . . on the nose at 15,000ft (4,570m)."

Hornet Two climbs to 25,000ft (7,620m) to gain an offensive position. Suddenly the radar blip separates.

"We've got two . . . the wingman is splitting high and left," replies Hornet Two. "I'm showing 1,000kt (1,840km/h) overtake."

Above: A Hornet of VFA-125 photographed at high altitude. The strangely foreshortened effect is a result of distortion caused by the canopy of the A-7 camera ship.

Below: Hornets of VMFA-314 Black Knights prepare to launch during their first sea deployment in July 1983. The Black Knights are assigned to CVW-13 aboard USS *Coral Sea*.

Above: In military power only, a clean Hornet of VMFA-323 Death Rattlers leaves the flight deck. The Hornet can be flown off the catapult 'hands-off'. VMFA-323 partners VMFA-314, VFA-131 and VFA-132 as part of the air wing embarked on *Coral Sea*.

Left: As a direct result of combat experience, pilots of modern fighters have an unobstructed view to the rear. The Hornet is comparable to any in this respect, as this view of a TF-18A back-seater shows.

Below: Pilot's eye view astern as a Hornet climbs gently away from the carrier it has just left in December 1982. It can be seen that blind spots, potentially so dangerous in air combat, have been reduced to a minimum.

"Six miles . . . in the box . . . four miles, tally-ho," replies Hornet One. "It's an A-4, shoot, shoot!"

"Fox One," calls Hornet Two, and the computer sends the simulated missile on its way to a lethal kill. As Hornet Two pitches back to assist, he hears Hornet One call a shot on the remaining bogey. The computer scores the shot as a miss as the bogey pilot hauls his aircraft into a vision-dimming 6g turn. Hornet Two manoeuvres into a cover position while Hornet One strives to regain the offensive advantage.

"Fox Two," calls Hornet One. "Good kill" is the call over the radio from the computer monitor station. Both aircraft turn west and 'bug out' in full afterburner, streaking towards 'good guy' country at the pass. The entire fight is over less than 30 seconds after the pilots initially sighted each other.

Pilots and aircraft

It may, of course, be argued that this exercise involved adversary aircraft which were far inferior to the Hornet and was therefore an unfair match. On the other hand, adversary pilots are men of high experience and ability levels, who do not do the job to be beaten and thus bolster their opponent's confidence; they play to win. Despite the discrepancy in the hardware, pilot ability historically always has been, and for the foreseeable future will continue to be the dominant factor in air fighting. Better technology helps a lot, but it is not the be-all and end-all.

In a contest between the Hornet and the Fighting Falcon, with pilots of equal ability, the result appears to be wide open. At the time of writing, the Hornet's weapon system is superior, and the Fighting Falcon has no effective answer to the 'shoot-in-the-face' capability conferred by the AIM-7 Sparrow. It then becomes a question of tactics. Initially, the Hornet pilot should try to keep his energy levels high and maintain his distance, taking Sparrow shots from any angle as opportunity offers, thus forcing the F-16 to manoeuvre hard to evade and deplete his energy. If this succeeded, the Hornet could then close for a heat missile or gun shot. In one versus one close combat, the contest would be fairly even.

It is the author's opinion that transient performance is more important than sustained turning ability. While no fighter in the world can match the sustained turn ability of the F-16, the pitch rate and the high AOA capability of the Hornet are believed to be better than those of the F-16, so that pilot ability would be the dominant factor in any contest between the two, and especially the ability of each pilot to use the strong points of his machine, which, rather surprisingly, is not always the case. For example, an Aggressor Squadron F-5E versus F-16 combat in 1982 saw the austere Northrop fighter achieve a good attacking position, whereupon the F-16 pilot evaded by a series of 9g loops which the F-5E was unable to match. The Aggressor pilot later commented: "I just flew around a while and waited for him to get tired!" This was not a good example of the use of sustained turn.

Multi-bogey combats

It must be remembered that one versus one combats are just peacetime training; they are not war. In war, multi-bogey combats are the norm, and they are, regardless of technical superiority, a great leveller. Leaving aside the BVR attack capability of the Hornet, one would expect the results of a multi-bogey contest between the F-16 and the Hornet to come out about equal, given equal numerical quantities. Given equal cost quantities, the cheaper F-16 would have a slight edge, but there are limits to the cost quantity equation. Given equal cost quantities, the austere F-5E is probably the greatest of them all, but procurement levels would be so high that pilot standards would have to drop to fill the empty cockpits. Moreover, the equal cost quantity argument, if carried to extremes, also involves the acceptance of an adverse kill ratio, which would be counterproductive due to the effect on morale. But to return to the Hornet, it may be fairly assumed that in the fighter role, it will perform extremely creditably.

The built-in reliability and maintainability of the Hornet paid handsome dividends during the second Yuma detachment. Originally, 288 sorties were planned, an average of 1.5 per aircraft per day. In the event, a total of 326 sorties exceeding 400 hours flight time were carried out, and every sortie had full systems operable. Only one sortie had to be cancelled, when a minor fault caused a 20-minute delay, which was enough to prevent the pilot from arriving at the range at his allotted time. Each Hornet averaged 44.7 hours of flight time during the deployment, and the remarkably low average of 11 MMH/FH was recorded. Not one fuel leak or hydraulic failure occurred during the deployment, a stark contrast to the Hornet's predecessor, the F-4 Phantom, of which it is often said, with some justification, that if it doesn't leak, it must be empty. The air combat training syllabus was satisfactorily finalized, and two months later a three-week deployment to NAS Fallon, in Nevada, served a similar function for the attack training syllabus.

Performance analysis

The original specification deficiencies either have been cured or do not seem to worry the pilots. The slow roll rate is now up to about 220deg/sec, about as much as a pilot can reasonably handle. The Hornet still does not accelerate from Mach 0.8 to Mach 1.6 in the required time, which was to give maximum energy at missile launch, but it is faster up to Mach 1.2 than virtually anything else, having beaten even the F-15 in 'drag races' up to this speed, and as hardly any air combat is likely to take place at speeds exceeding Mach 1.2, the pilots consider this deficiency to be unimportant.

The Hornet has attracted a lot of flak over its range capability, but many of the comparisons drawn seem suspect, especially those involving the Corsair. Fuel flow management and close attention to flight profiles for the attack mission have produced improvements, and although a Hornet in the attack configuration does not have the radius of action of the Corsair that it replaces, the difference is reportedly less than 10 per cent. The Hornet's far greater survivability due to its vastly superior performance is considered by the pilots to be more than an adequate trade-off for a reduction in maximum range.

From the staff point of view, an operational evaluation in 1982 stated that the unrefuelled capabilities of the Hornet cause a reduction in the stand-off range of a battle group, an important consideration in the event of an action against a Soviet fleet equipped with SS-N-12 or SS-N-19 surface-launched anti-shipping missiles with a range of 300nm (550km). Nevertheless, it appears that the criteria for the Hornet should not be performance comparisons against other types, but how well it performs its allotted role.

In the final analysis, theoretical ranges attainable at certain speeds under precise conditions with defined weapon loads are irrelevant. With the attack role specifically in mind, the only question worth asking is, can the Hornet deliver a worthwhile load at a sufficient distance (not necessarily greater in either case) with greater accuracy and a greater margin of survivability than the aircraft that it is to replace? In the air combat role, the question becomes, can the Hornet perform both as an interceptor and close combat fighter in an effective manner without being outclassed by the opposition? In both cases, the answer must be an unqualified yes. The pilots genuinely praise the Hornet's handling qualities, although with neutral speed stability the control column does not have a lot of 'feel', and a certain amount of care must be taken to avoid overstressing the airframe. Maximum lift, achieved at about 35deg AOA, can be reached with a one-handed pull on the stick, while to reach higher AOA two hands are needed. The aircraft is almost spinproof, while if it does depart controlled flight, recovery is simple. It is stable in both roll and yaw, and is a good gun platform up to 30deg AOA.

Carrier qualification

A further milestone in finalizing the Hornet training syllabus was reached between September 27 and October 4, 1982, when the Rough Raiders completed their first carrier qualification operations aboard the USS *Constellation*. Six pilots using two single-seat Hornets logged ten bolters as well as 57 day and 24 night traps during this period, and no problems were encountered.

Catapult launches are regarded as easy. The Hornet taxis to the cat, and the launch bar, which is attached to the nose gear, is positioned in the shuttle on the catapult track. The pilot makes his final checks, then waves to the catapult officer that he is ready to go. The Hornet is trimmed to fly 'hands off', though it is normal for the pilot to put his left hand on the throttle to prevent it slipping back under the 4g acceleration as the catapult fires.

Landing is carried out either manually or automatically. For a manual landing the pilot sets up the speed, line-up and glideslope. 'On-speed' is taken from the aircraft's instruments, while line-up is judged from the centreline on the carrier

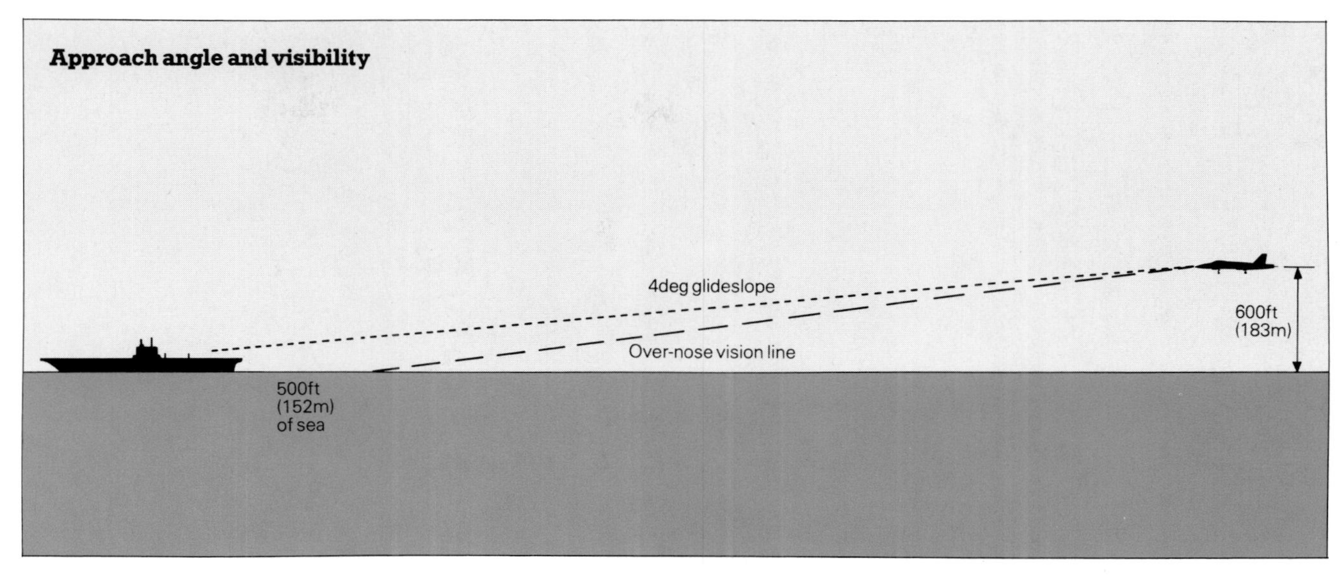

Above: The Hornet has a low AOA on the final approach, and the excellent visibility over the nose simplifies landing on a carrier deck to a considerable degree.

Below: The carrier suitability trials Hornet launches from a 6deg ski-jump ramp at NATC Patuxent River. The attitude of the main gear suggests that a fair proportion of the aircraft's weight, though not yet all of it, has been transferred to the wings.

deck. The glideslope, an angle of about 4deg, is taken from the Fresnel Lens, which is a series of directional lights on the carrier: the pilot watches the mirror of the Fresnel Lens and the correctness of his approach is shown by the colour of the light reflected in the mirror, an orange light indicating that the glide-slope is correct. At touch-down the throttles are advanced to full military power; if the hook misses the wires the Hornet flies right on past the carrier and tries again.

The automatic system is different again. At a set distance out on the approach the pilot couples his ACLS to the ACLS SPN-42 system on the carrier. This transmits signals to the Hornet's auto-pilot which guide the aircraft straight on to the carrier deck, making due allowance for deck motion. The throttles are controlled automatically in this mode by the Hornet's approach power compensator (APC).

Training syllabus

The pilot training syllabus evolved by VFA-125 lasts five months, a little longer than the course for maintenance personnel, and consists of four phases: transition, air-to-air, air-to-ground, and carrier qualification. Instruction begins in the brand new Hornet Learning Center at Lemoore, where computer-assisted

Left: Final checks are carried out as one of the VMFA-314 Hornets prepares to launch from the catapult aboard USS *Constellation* during carrier operations in July 1983.

Below: Deck crew cluster round Hornets of VMFA-314 as they get ready to launch. The coloured vests indicate the various functions to which the individual crewmen are assigned – purple and a white 'F' identify the fuelling team.

audio-visual techniques cover procedures, switchology and systems operation, and the fledgling Hornet pilot then progresses to the simulators. The first of these is the Part-Task Trainer (PTT), consisting of simplified Hornet cockpits, in which familiarization with HOTAS is begun and the manipulative skills needed to operate the controls is acquired. Then comes the Operational Flight Trainer (OFT), which simulates actual flight from takeoff to landing by means of three visual displays and uses a g-suit/g-seat buffet system for extra realism. OFT training covers both airfield and carrier takeoffs and landings, the complete instrument syllabus, and emergency procedures.

Then comes the Weapons Tactics Trainer (WTT), consisting of two 40ft (12.2m) domes with a cockpit in each and seven televisual projectors. The WTT, in which each student will spend an estimated 50 hours, provides advanced air-to-air radar training and ACM practice. Each dome can be operated solo, or combined into a 'twin-tub' to allow the pilots in each dome to 'fly' against each other. The projectors generate a realistic earth/sky backdrop on the inside of the domes on which target, missile and gunfire images are also generated. After sufficient time has been spent in the simulators, flight training begins. This consists of around 70 sorties, supplemented by the simulators as necessary, followed by a further 20 sorties for carrier qualification.

Operational squadrons

The first operational Hornet squadron was VMFA-314, the Black Knights, commanded by Lt. Col. Peter Field, who had played a major part in the Hornet FSD programme at Patuxent River. The Black Knights were part of the Phantom-equipped 3rd Marine Air Wing based at MCAS El Toro, California, and the inaugural ceremony took place on January 7, 1983, at El Toro. At a press conference immediately afterwards, Lt. Col. Field had a number of points to make about the Hornet in reply to questions from reporters.

The first concerned the accuracy of weapons delivery, which was described as better than ever before, with a weapon system "literally out of Star Wars". The second point concerned the sheer pleasure of flying the Hornet. This contrasts with the Phantom, which has never been regarded as an easy aeroplane to fly and requires a lot of concentration. The third point concerned the maintainability of the Hornet, with MMH/FH reduced to one-third of the figure for the Phantom. He commented that the squadron had achieved more flight hours over a given period with their first four Hornets than they had been used to getting from their full complement of 12 Phantoms. Later in the year, the Black Knights participated in Red Flag 83/5 at Nellis AFB, where they flew as aggressors, clashing with Air Force F-15s and F-16s. Unfortunately, the details are still classified at the time of writing.

The Black Knights were followed on to the Hornet by their fellow squadrons of the 3rd MAW, VMFA-323 and VMFA-531, and the Navy was not far behind, VA-113 Stingers commencing training for the Hornet on March 26, 1983. They became operational in August of the same year, a full two months ahead of schedule. The Stingers were previously equipped with Corsairs, and their commanding officer, Cdr. William Pickavance, was quick to comment that for years the squadron had needed fighter escort, whereas they now had an aircraft with an excellent self-defence capability.

Hornet squadrons

US Navy and Marine Corps F/A-18 Hornet squadrons as of May 31, 1984

Unit	Tailcode	Name and base	Carrier assignment	Remarks
Operational Test and Evaluation				
NATC	7T	(none) NAS Patuxent River, Md	(none)	From 1978, operational tests.
NWC	(none)	(none) NWC China Lake, Ca	(none)	From 1984, weapons tests.
VX-4	XF	Evaluators NAS Point Mugu, Ca	(none)	From 1981, operational test and evaluation; carrier trials with CVW-15/USS *Vinson* Fall 1982; carrier trials with VFA-125 on USS *Constellation* Fall 1982; CO Capt J. Michael Welch.
VX-5	XE	Vampires NAS Point Mugu, Ca	(none)	From 1981, operational test and evaluation; carrier trials with CVW-15/USS *Vinson* Fall 1982; CO Cdr Michael P. Curphey.
Pacific Fleet Replacement Air Group (RAG)				
VFA-125	NJ	Rough Raiders NAS Lemoore, Ca	(none)	Commissioned Nov 13, 1980, first F-18 squadron; received first aircraft Feb 19, 1981; ACM deployment to MCAS Yuma, Az, Jan 5-27, 1982; bombing deployment to NAS Fallon, Nev, Mar-Apr 1982; carrier trials with VX-4 on USS *Constellation* Sep-Oct 1982; CO Capt James W. Partington.
Pacific Fleet				
VFA-25	NK	Fist of the Fleet NAS Lemoore, Ca	CVW-14 USS *Constellation*	Commissioned Jul 1, 1983; received first aircraft Nov 11, 1983; to deploy with VFA-113.
VFA-113	NK	Stingers NAS Lemoore, Ca	CVW-14 USS *Constellation*	Commissioned Mar 25, 1983; received first aircraft Aug 16, 1983; to deploy with VFA-25; CO Cdr Craig A. Langbehm.
Atlantic Fleet Replacement Air Group (RAG)				
VFA-106	AD	Gladiators NAS Cecil Field, Fla	(none)	Commissioned Apr 27, 1984; CO Cdr David J. L'Herault.
Atlantic Fleet				
VFA-131	AK	Wildcats NAS Cecil Field, Fla	CVW-13 USS *Coral Sea*	Commissioned Mar 29, 1984.
VFA-132	AK	Privateers NAS Cecil Field, Fla	CVW-13 USS *Coral Sea*	Commissioned Feb 10, 1984; CO Cdr Robert E. Lakakri.
VMFA-314	VW	Black Knights MCAS El Toro, Ca	CVW-13 USS *Coral Sea*	Converted from F-4N Jan 7, 1984; assigned to CVW-13 Mar 29, 1984; CO Lt Col Pete Field.
VMFA-323	WS	Death Rattlers MCAS El Toro, Ca	CVW-13 USS *Coral Sea*	Converted from F-4N 1983; assigned to CVW-13 Mar 29, 1984; CO Lt Col Gary Vangysel.
Other units				
VMFA-531	EC	Gray Ghosts MCAS El Toro, Ca	MAG-11, 3rd MAW	Converted from F-4N 1983; CO Lt Col Jim Lucas.
VFA-136	AK	(unassigned) NAS Cecil Field, Fla	CVW-13 USS *Coral Sea*	To be commissioned 1985; with VFA-137 to replace VMFA-314 and VMFA-323 aboard *Coral Sea*.
VFA-137	AK	(unassigned) NAS Cecil Field, Fla	CVW-13 USS *Coral Sea*	To be commissioned 1985; with VFA-136 to replace VMFA-314 and VMFA-323 aboard *Coral Sea*.
VFA-146	(unassigned)	(unassigned) NAS Lemoore, Ca	(unassigned)	To convert from A-7E about 1986.
VFA-147	(unassigned)	(unassigned) NAS Lemoore, Ca	(unassigned)	To convert from A-7E about 1986.
VFA-303	ND	Golden Hawks NAS Miramar, Ca	CVW-30 Naval Air Reserve	To convert from A-7E about 1986 as first Reserve F-18 squadron.

NOTE: *Coral Sea* is currently being refurbished and will be the first carrier with four F/A-18 squadrons (replacing two A-7E and two F-4S squadrons); the assignment of Marine Corps squadrons is temporary pending the formation of VFA-136 and VFA-137.

CO: Commanding Officer **CVW:** Carrier Air Wing **MAG:** Marine Aircraft Group **MAW:** Marine Aircraft Wing **NATC:** Naval Air Test Center **NWC:** Naval Weapons Center **VFA:** (US Navy) Strike Fighter Squadron (formerly Fighter Attack Squadron) **VMFA:** Marine Fighter Attack Squadron **VX:** Operational Test and Evaluation Squadron

LOCATIONS: Az Arizona **Ca** California **Fla** Florida **Md** Maryland **Nev** Nevada

Two of the biggest controversies about the Hornet appear to have been largely silenced since its operational debut. The first concerned the multi-role capability insofar as it affected the pilot. Some pilots are natural air fighters, while others really enjoy moving mud. The opinion has been widely held that never the twain should meet, and while units had specialist roles, pilots tended to be assigned to squadrons where their natural bent could best be utilized. Reports from the squadrons appear to indicate that interchangeability appears to be working, although doubts will continue to exist for some time yet. Of course, a squadron contains more pilots than aircraft, so it seems logical that if the split between natural fighter and natural attack pilots is about even, then the nature of the mission can be allowed to affect the allocation of the pilots within certain limits.

Solo mission capability

The other doubtful area has been whether one man can fly the type of mission that has come to be regarded as the prerogative of the two-seater. The general opinion of those who fly the Hornet appears to be that the advanced avionics, and especially the cockpit, enable one man to perform satisfactorily, although this verdict is not unanimous. Some former Skyhawk and Corsair pilots would prefer a back-seater to help with the workload, despite, or perhaps because of their previous experience. It is, however, accepted in the Hornet squadrons that this is a minority opinion. As the old saying goes, "Combat is the ultimate, and the unkindest, judge."

At least 40 USN and USMC squadrons are planned, although this number, as well as the total planned procurement of the Hornet, may change if Navy and Marine Corps mission requirements are revised. On the large carriers, the Hornet will supplement the Tomcat in the fleet defence role, but it is planned that the two smaller carriers, USS *Midway* and *Coral Sea*, which will both remain in service into the 1990s, will deploy four squadrons of Hornets each.

Above: Two CF-18B two-seaters of the CF take on fuel from a CC-137 tanker. Flight refuelling is essential for the Canadian Forces, given the vast size of the country. Dummy canopies are painted on the undersides of all Canadian CF-18s for aspect deception of opponents in combat.

The excellent high AOA handling of the Hornet, coupled with its high thrust to weight ratio, have led McDonnell Douglas to suggest that it might be capable of a ski-jump takeoff in the manner of the Harrier. Tests are currently underway at Patuxent River using 6deg and 9deg ramps, and it has been suggested that a 25deg ramp may eventually be tried. As the Hornet does not have vectored thrust, the full operational implications are at present unclear, although two-dimensional thrust vectoring has been suggested for the F404 engine. The great advantage of a ramp takeoff for shipboard operations is that no matter how heavy the swell, the aircraft is launched upwards away from the sea. But whether a combined ramp and catapult is intended, or even feasible, remains to be seen.

Many air forces have evaluated the Hornet with a view to re-equipping with the type, and to date Canada, Australia, and Spain have placed firm orders. One of its main competitors has been the Hornet's original rival, the General Dynamics F-16, which has been very successful in securing overseas orders. The two main criteria appear to be cost and mission effectiveness: where the F-16 can perform the required missions it is definitely an attractive proposition due to its lower cost, whereas the more capable Hornet is able to meet more stringent operational specifications but at greater cost.

Export advantages

Nevertheless, the F-16 drew considerable flak in Europe when it first entered service for its lack of adverse weather BVR capability, and its limited armament. The original two Sidewinders would not last long in a major air battle, and the consequent lack of on-board kills was likely to seriously limit its combat persistence. The F-16 is also stuffed full of fuel in a manner that arguably increases its vulnerability to battle damage. By contrast, the Hornet has a genuine adverse weather capability and was equipped with BVR weaponry from the outset. In the fighter configuration, it has never had less than four missiles, and can easily carry more. Its two engines give greater flight safety, and it has exceptional built-in survivability.

Canada is a country of vast size with a sparse population over most of its area. It also has air defence commitments in Europe, and the McDonnell Douglas CF-101 Voodoos, Lockheed CF-104 Starfighters, and Northrop CF-5 Freedom Fighters operated by the Canadian Armed Forces were looking distinctly long in the tooth by the end of the 1970s. After an intensive evaluation of many contending fighters, the Canadian government became the first export customer for the Hornet, signing a contract for 137 aircraft, including 24 two-seaters, in the summer of 1980 later adding another 11 single seaters.

The trials had ended in a straight contest between the F-16 and the Hornet, and two key factors were cited in the decision to adopt the Hornet rather than the F-16. The first was twin-engined safety. Flying over Canada has certain similarities to flying over the sea; airfields are few and far between, and a pilot forced to eject could be lost almost as easily in the inhospitable wilderness as over water. The second reason was that the Canadian authorities felt that the larger F/A-18 had more growth potential than its single-engined rival.

The Canadian Hornet features three changes from its US Navy counterpart.

Right: the Royal Australian Air Force has ordered a total of 75 Hornets to replace three squadrons of Mirage IIIs. Among the distinctive features of the RAAF Hornets are the IFF antennae under the nose.

On the left side of the forward fuselage is mounted a 600,000-candlepower spotlight, a standard Canadian item used to make visual identification of bogeys at night. A different ILS is fitted, and in lieu of the USN sea survival gear, a cold-weather land survival kit is carried.

Canadian camouflage

A further difference, though hardly a modification, is the painting of a false canopy on the underside of the nose. The brainchild (and also patent) of American aviation artist Keith Ferris, it is intended to deceive opponents as to which way the aircraft is turning. Extensive experiments have been carried out in the United States with this and other forms of deception camouflage, but with inconclusive results. It obviously works some of the time, as complaints have been received about it being a potential collision hazard in peacetime training exercises, although it cannot work at very close ranges nor yet at extreme visual distances. In wartime it could have a certain value, as it interferes with what former US fighter instructor Major John Boyd calls the opposing pilot's OODA loop (Observation, Orientation, Decision, Action). The deception interferes, even if only briefly, with the orientation phase, thus delaying the decision and action phases, which in a time-critical situation could be valuable indeed. The false canopy is generally painted matt black with irregular gloss black splodges to represent the glint of sunshine on a canopy.

The first two CF-18s – the name Hornet has not been adopted by the Canadian Armed Forces – CF 901 and 902, were delivered to the Aerospace Engineering Test Establishment at Cold Lake, Alberta, on October 25, 1982, and the first CF-18 unit to be formed was No. 410 Operational Training Squadron, whose pilots had been through the VFA-125 training syllabus. The three operational squadrons to be formed will be Nos. 421 and 441 Squadrons, while in June 1985, the Hornet will be introduced to European service by No. 439 Squadron, presently flying Starfighters out of Soellingen in West Germany.

Above: The first Canadian Forces unit to be formed was No. 410 Operational Training Squadron at CFB Bagotville, Quebec, to which this gleaming new two-seater is assigned. Unit badge is a cougar head.

Below: The first two CF-18s to be delivered fly in formation. The 600,000-candlepower spotlights carried by Canadian aircraft can be seen ahead of and below the LEX.

Australian Hornets

In many ways, the problems of the Royal Australian Air Force duplicate those of the US Navy and Canadian Armed Forces. Much of the RAAF's flying will be done over water; sea surveillance is high on its list of priorities; and like Canada, Australia is a large, sparsely populated and inhospitable country. In a hard-fought contest against the F-16, the Hornet was selected for three main reasons, principally for its twin-engined safety and better growth potential; thirdly, it was decided that the proposed up-grading of the F-16's avionics, in particular the radar, all-weather targeting system and navigation equipment, presented a risk, whereas the Hornet systems were adequate without alteration. A total of 75 aircraft, of which 18 will be two-seaters, were ordered to replace the three squadrons of ageing Dassault Mirages.

The Australian Hornet will differ from the standard in minor details. They include the elimination of catapult launch equipment, the replacement of carrier ILS by a conventional ILS, and the provision of HF radio for long distance communications, indigenous fatigue monitoring, Tacan and IFF systems, a landing light, and an aural 'gear down' warning system. The first Australian Hornet components reached the Saint Louis assembly line in August 1983, and handover of the first two aircraft is expected to take place in October 1984. These will then be used to train OCU pilots before being flown to Australia in April 1985. The third Hornet will be supplied in knock-down kit form for re-

assembly by the Government Aircraft Factory at Avalon, which will be responsible for the assembly of the remaining 72 aircraft. The F404 engines for the Hornet will also be assembled in Australia, by Commonwealth Aircraft at Melbourne. The Australian-assembled Hornets are scheduled to commence delivery from April 1985.

A major construction programme is currently underway at Williamtown in New South Wales involving hangarage, a maintenance complex, a simulator building, and other facilities for Hornet operations. In addition, major works are taking place at Tyndal in the Northern Territory, some 300 miles (480km) south of Darwin, which will be another Hornet's Nest. The first operational unit to receive Hornets will probably be No. 75 Squadron, currently based at Darwin, followed by Nos. 3 and 77 Squadrons. It has also been proposed to establish another base at Weipa, in northern Queensland, which would bring Papua, New Guinea, within range of the Hornet. One additional and not inconsiderable advantage of the Hornet in RAAF service may accrue from the fact that US Navy carrier groups operating in the Pacific and Indian oceans would be using the same type, thus facilitating mutual assistance.

Spanish Hornets

The third nation to order the Hornet was Spain, with the F-16 being the final alternative once more. The aircraft to be replaced were F-4C Phantoms (with which the Ejercito del Aire had never been very happy), Mirages and Freedom Fighters. Originally 144 aircraft were required, but budgetary considerations saw this reduced to 72 Hornets, plus an option on a further 12, and the contract was signed on May 31, 1983, with first delivery due in 1986.

The Spanish operational requirement was for an adverse-weather, day or night aircraft able to operate in either the fighter or attack roles, although attack was given the greater emphasis. Recent head-to-head sales contests against the F-16, notably those involving Greece and Turkey, as in many previous cases, have been lost to the cheaper single-engined fighter. Yet it has been noticeable that whenever operational requirements were more demanding, the Hornet has been the aircraft selected.

The Luftwaffe is currently evaluating the Hornet, together with the McDonnell Douglas F-15, the General Dynamics F-16, the Northrop F-20 Tigershark, and the Dassault Mirage 2000. Air superiority is the prime requirement, with a procurement of 200 to 300 aircraft at stake. If capability regardless of cost is the aim, the F-15 Eagle has to be the favourite, whereas if cost is the main consideration, the Tigershark should win hands down. If maximum performance in the top right-hand corner of the performance envelope is required, the Mirage 2000 cannot be ruled out. But for a good all-rounder at an in-between price, the contest is likely to devolve upon those old adversaries, the Fighting Falcon and the Hornet.

In fact, it is widely believed that the West German evaluation is being performed in order to establish a datum for the proposed TKF fighter specification. But this being so, the development period of a totally new fighter is such that the existing F-4F Phantom fleet will reach the end of its effective life before the new fighter is ready, in which case the Hornet, with its all-round capability, would make a more than useful stopgap. It appears to be the aircraft to beat.

The Hornet, like most combat aircraft, will doubtless see service in many specialized roles during its lifetime, and one of these is already emerging in the shape of the reconnaissance RF/A-18. It was originally planned to produce a dedicated two-seat variant for this role, but studies showed that extensive structural changes would be needed, as well as a new environmental control system to cool the additional avionics.

Reconnaissance conversion

Consequently, it was decided to produce a reconnaissance pallet which would be interchangeable with the Vulcan cannon. The gun cavity door on the underside of the nose was also to be replaced by a bulged fairing with viewing ports and provision for extra sensor mountings, and the entire package was designed to be installed or replaced at squadron level in about eight hours. Testing began in October 1982 and was scheduled to continue until September 1984, but the package concept, like many another lash-up before it, proved unsatisfactory, and early in 1984 the US Navy reverted to its original scheme and awarded a development contract to McDonnell Douglas to produce a dedicated reconnaissance Hornet. The gun is to be removed and cameras installed in its place, while heating and cooling lines are to be modified, and the software is to be reprogrammed to suit the sensors. Flight testing is scheduled to commence in the summer of 1984 and be completed sometime during the following year. It is possible that extra fuel will be accommodated by raising the 'spine' of the aircraft behind the cockpit.

Another possible variant that has been under discussion for some time is the two-seat heavy attack aircraft to replace the A-6E Intruder. No designation has

Above: A considerable effort was made to sell the Hornet overseas. This 1980 photograph shows an early prototype, with unfilled LEX slots and notched wing leading edge, in Swedish Air Force markings.

Below: A reconnaissance Hornet is the next step, and MCAIR received a development contract in early 1984. Sensors will replace the gun, the 1982 proposal for pallet-mounted equipment having been rejected.

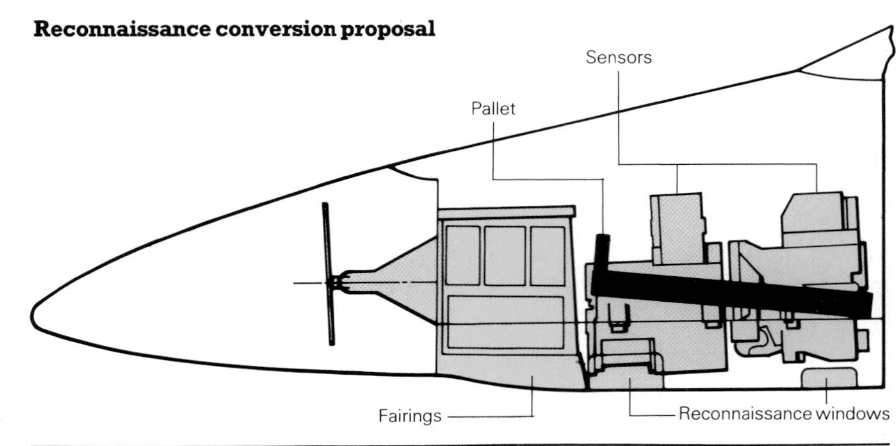

Right: Spain was the third export customer for the Hornet, the 72 ordered being due for delivery from 1986. By mid-1984 no decision had been made on colour schemes for the Ejercito del Aire, but this illustration shows how an F-18 might appear in similar camouflage to that worn by Spanish Air Force Phantoms.

Left: Mock-up of the two-seat Northrop F/A-18L. Sparrows occupy the wingtip launch rails, while Sidewinders are carried on extra underwing pylons. The Northrop proposal has lost out to the F-16 and F-18A in terms of sales.

been assigned as yet, but it is unofficially referred to as the Advanced A-18. A McDonnell Douglas proposal incorporates modifications to the trailing edges and flaps, uprated F404 engines in the 18,000lb (8,165kg) thrust class and expanded air-to-surface radar modes based on experience with the Strike Eagle programme. Additional internal fuel capacity, a mission-adapted rear cockpit and larger external fuel tanks are also proposed.

Northrop F-18L

As related in the development chapter, McDonnell Douglas took the Northrop YF-17 lightweight fighter and upgraded the design into a carrier-compatible multi-role fighter, the F/A-18A. At the same time, Northrop produced a design for a land-based multi-role variant with about 90 per cent commonality in terms of high-value, high-usage parts, which was originally intended to be about 20 per cent cheaper than the F/A-18A carrier fighter, but with the same versatility. It was designated F/A-18L.

Not being intended for carrier operations, the Northrop Hornet has a simpler and much lighter undercarriage, stressed for descent rates of up to 14fps (4.27m/sec). This allows greater external stores carriage on the fuselage, and also leaves an extra 18cu ft (0.51m^3) available for extra avionics and fuel. A smaller and lighter hook is fitted for runway arrestment, and the fuselage missile housings of the F/A-18A are not featured, thus saving weight and drag; Sparrow or Skyflash missiles are carried on underwing pylons, or even on the wingtips in lieu of Sidewinders.

There is no necessity for the wings to fold and the hinges and hydraulic folding gear have been eliminated. This has allowed the wings to be strengthened to carry the much heavier SARH missiles on the tips, and an extra hardpoint has been included on each wing. Instead of the ailerons and flaps on the F/A-18A, flaperons are used which incorporate ride control by varying the lift and apparent AOA to reduce the gust response, thus giving a smoother ride in high-speed low-altitude flight. On touchdown the flaperons deflect vertically and increase the download on the undercarriage, which makes the brakes more effective, and reduces the landing run. The flap deflection on the F/A-18L varies from its naval counterpart. The maximum flap deflection is 20deg for the leading edge flaps and 30deg for the trailing edge, compared with 30deg and 40deg respectively for the F/A-18A. Integral wing tanks are an optional extra, while USAF type 600US gall (1,756lit) drop tanks can be carried on the inner underwing hardpoints.

The overall effect of these changes is to reduce the aircraft weight by nearly a tonne. Thus wing loading is reduced, which improves manoeuvrability, and thrust loading is increased, which benefits acceleration, takeoff, climb and sustained turn. Alternatively, a heavier payload can be carried: the stated maximum is 20,000lb (9,072kg), which is rather more than the carrier fighter. With these advantages, it would seem that if the F/A-18A is a handful in the dogfight, the F/A-18L would be even more so. Yet no orders have been placed, and the F/A-18L bids fair to become the fighter that never was. The two YF-17s were judged sufficiently similar to be redesignated F/A-18L prototypes by Northrop for evaluation purposes, but not one F/A-18L has been built, and it is beginning to look as though none ever will be. This is the more surprising as the market potential was originally assessed as being around the 2,000 mark.

Part of the problem has been that the F-16 has been in competition, and the F-16 is a very fine fighter indeed. The other part arises from the fact that McDonnell Douglas have been energetically marketing their Hornet abroad, often in competition with the Northrop Hornet. Evaluation teams can go to Fort Worth or Saint Louis and see a finished product, whereas at Hawthorne this is not possible. One feels that Northrop and the F/A-18L Hornet deserved better. In early 1984 lawsuits pending between the rival companies were alleging what is in essence to be described as poaching. The legal situation is complex, involves Department of Defense rulings and is beyond the scope of this work. Although part of the Hornet story, with a decision anticipated late in 1984, it is of little interest from a strictly aviation viewpoint. What does matter is that the Hornet is a very fine warplane and a worthy successor to the ubiquitous Phantom.

Glossary and abbreviations

A&AEE Aircraft and Armament Experimental Establishment (UK)
ACF Air Combat Fighter (USAF)
ACLS Automatic carrier landing system
ACM Air combat manoeuvring
ADF Automatic direction finding
AFB Air Force Base (USAF)
AGM Air-to-ground missile
AIM Air interception missile
AIMVAL/ACEVAL Missile and air combat evaluation programme carried out in Nevada in 1977
Alpha-numeric Information presented in the form of letters and/or numbers
AMAD Airframe mounted accessory drive
AMRAAM Advanced Medium-range Air-to-air Missile
AMT Accelerated mission test
Analogue Electronic system in which quantities are represented by electrical signals of variable characteristics, i.e., by electrical analogues
Anhedral Downward angle of wing or tailplane
AOA Angle of attack (the angle at which the wing meets the airflow)
APC Approach power compensator
APU Auxiliary power unit
Aspect ratio Ratio of the wingspan to the wing area, expressed as span squared divided by wing area
AST Accelerated service testing
Azimuth Bearing or direction in the horizontal plane
BIT(E) Built-in test (equipment)
Bolter Touch-and-go deck landing (usually unintentional)
Boundary layer Thin layer of slow-moving air that tends to cling to the skin of an aircraft
Bug out Depart the area
Bu. No. Bureau of Aeronautics number
BVR Beyond visual range
Bypass ratio Ratio of total volume of air passing through the engine to that passing through the core section.
CNI Communications, navigation, identification
CO_2 Carbon dioxide
CRT Cathode ray tube
CW Continuous wave (radar emission)
DBS Doppler beam sharpening
Departure The point at which an aircraft goes out of control
Dielectric Radar non-reflecting material
Digital Electronic system in which quantities are represented as on/off signals coded to represent numbers
DoD Department of Defense (US)
Doppler Radar making use of shift in frequency of signals reflected from the earth's surface ahead of or behind an aircraft to give measurement of true groundspeed, or of signals reflected from earth and moving targets to indicate the latter
ECCM Electronic counter-countermeasures
ECM Electronic countermeasures
FBW Fly-by-wire (electronic flight control system)
FET Field effect transistor
Flaperon Control surface doubling as flap and aileron
FLIR Forward-looking infra-red
FM Frequency modulated
FOD Foreign object damage
Fox One Pilot call on launching a Sparrow
Fox Two Pilot call on launching a Sidewinder
FSD Full scale development
g Unit of acceleration
GHz Giga Hertz; 1,000,000,000 Hertz (cycles per second)
Glint Apparent movement of the radar centre of a target
HDD Head-down display
HF High frequency
HOTAS Hands on throttle and stick
HP High pressure
HSD Horizontal situation display
HUD Head-up display
IFECMS In-flight engine condition monitoring system
IFF Identification friend or foe
IFR In-flight refuelling
IIR Imaging infra-red
INS Inertial navigation system
IR Infra-red
Knot Nautical mile per hour
kVA kiloVolt Amperes
LEX Leading edge extension (at the wing root)
LGB Laser-guided bomb
Lock-on Radar concentrating on targets in the attack mode
LP Low pressure
LRU Line replaceable unit
LWF Lightweight Fighter (USAF)
MAP Military Assistance Programme
Mach number Speed stated as a function of the local speed of sound
MCAS Marine Corps Air Station
MER Multiple ejection rack
MFD Multi-function display
MFHBF Mean flight hours between failures
MMD Master monitor display
MMH/FH Mean maintenance hours per flight hour
MMP Maintenance monitor panel
MTBF Mean time between failures
NAS Naval Air Station
NATC Naval Air Test Center
Nm Nautical mile (= 1.15 statute miles, 1.85km)
NPE Naval Preliminary Evaluation
OFT Operational Flight Trainer
OODA loop Sequence of pilot's mental processes: observation, orientation, decision, action
Overtake Closing speed irrespective of relative aspect or heading
Passive Non-emitting
Pitch Vertical movement or angle of aircraft longitudinal axis
PPFRT Prototype Preliminary Flight Rating Test
PRF Pulse repitition frequency
Ps Specific excess power
PSP Programmable signal processor
PTT Part Task Trainer
Raster Television picture built up line by line
Red Flag Tactical exercises held at Nellis AFB
RTS Radar Test System
RWR Radar warning receiver
SARH Semi-active radar homing
Sfc Specific fuel consumption (unit of fuel consumed per unit of thrust per hour)
SMET Simulated mission endurance test
Stabilators All-moving one-piece tailplane
STT Single target track
Tacan Tactical air navigation
TACTS/ACMI Tactical Air Combat Training System/Air Combat Manoeuvre Instrumentation
Taileron All-moving tailplanes with differential movement to provide control in the rolling plane
TDC Target designator control
TER Triple ejection rack
TFTS Tactical Fighter Training Squadron
TFW Tactical Fighter Wing
Trap Arrested deck landing
TWT Travelling wave tube
UFC Up-front control
UHF Ultra-high frequency
US gall US gallon (= 0.83Imp gall; 3.785lit; 6.5lb [2.95kg] JP-4 fuel)
VDU Visual display unit
VER Vertical ejection rack
VHF Very high frequency
Vmax Maximum velocity
WRA Weapon replaceable assembly

Specification

	YF-17	F-18A	F-18L
Dimensions			
Span (without missiles)	35ft 0in (10.67m)	37ft 6in (11.43m)	37ft 6in (11.43m)
Length	56ft 0in (17.07m)	56ft 0in (17.07m)	56ft 0in (17.07m)
Height	14ft 6in (4.42m)	15ft 3½in (4.66m)	14ft 7in (4.44m)
Tail span	22ft 2½in (6.77m)	21ft 7¼in (6.58m)	21ft 7¼in (6.58m)
Wheel track	6ft 10in (2.08m)	10ft 2½in (3.11m)	8ft 5in (2.57m)
Wing area	350sq ft (32.52m^2)	400sq ft (37.16m^2)	400sq ft (37.16m^2)
Weights			
Empty	17,000lb (7,700kg) approx	21,830lb (9,900kg)	19,600lb (8,900kg) approx
Fighter configuration	23,000lb (10,430kg)	34,700lb (15,740kg)	32,000lb (14,500kg) approx
Maximum		51,900lb (23,540kg)	
Performance			
Vmax	Mach 1.95	Mach 1.8	Mach 2+
Combat ceiling	50,000ft (15,240m)	50,000ft (15,240m)	55,000ft (16,765m)
Combat radius – fighter	>500nm (927km)	>400nm (740km)	
Combat radius – attack		575nm (1,065km)	
Ferry range (unrefuelled)	2,600nm (4,816km)	>2,000nm (3,706km)	2,500nm (4,630km)
Initial climb rate	>50,000ft/min (254m/sec)	50,000ft/sec (254m/sec)	56,000ft/sec (285m/sec)
Maximum payload	9,800lb(4,445kg) approx	17,000lb (7,710kg)	20,000lb (9,070kg)

Picture credits

Endpapers: McDonnell Douglas. **Title page:** McDonnell Douglas. Page **2/3:** McDonnell Douglas. **4:** (both) Northrop. **5:** (top) Northrop; (bottom) USAF. **6/7:** (all) Northrop. **8/9:** (all) McDonnell Douglas. **10/11:** (all) McDonnell Douglas. **12:** (top) Robert F. Dorr; (centre and bottom) McDonnell Douglas. **13:** (top) Robert F Dorr; (centre) McDonnell Douglas. **14/15:** (all) McDonnell Douglas. **16/17:** (all) McDonnell Douglas. **18/19:** (all) McDonnell Douglas. **20/21:** (all) McDonnell Douglas. **22/23:** (all) McDonnell Douglas. **24:** McDonnell Douglas. **25:** (top) Northrop; (centre) McDonnell Douglas. **26/27:** (all) General Electric. **28:** McDonnell Douglas. **29:** US Navy. **30/31:** (all) McDonnell Douglas. **32:** (all) McDonnell Douglas. **34:** (top) Kaiser Electronics; (bottom) Ferranti. **36:** (centre) Hughes Aircraft; (bottom) Robert F. Dorr. **37:** (all) Hughes Aircraft. **40:** (top) Sperry Corporation; (bottom) McDonnell Douglas. **41:** (top) McDonnell Douglas; (bottom) Goodyear Aerospace. **42/43:** (all) McDonnell Douglas. **44:** (top) McDonnell Douglas; (bottom) Canadian Forces. **45:** McDonnell Douglas. **46:** (top) McDonnell Douglas; (centre) General Electric. **47:** (both) McDonnell Douglas. **48/49:** (all) McDonnell Douglas, **50:** Canadian Forces. **51:** (top and centre) Hughes Aircraft; (bottom) Canadian Forces. **52:** (centre) US Navy; (bottom) McDonnell Douglas. **54:** (both) McDonnell Douglas. **55:** (top) US Navy; (bottom) McDonnell Douglas. **56:** (all) McDonnell Douglas. **57:** US Navy. **58:** (both) McDonnell Douglas. **60:** Canadian Forces. **61:** (top) Robert F. Dorr; (centre) Canadian Forces. **62:** McDonnell Douglas. **63:** Northrop.

PRINTED IN BELGIUM BY proost INTERNATIONAL BOOK PRODUCTION